WITHDRAWN
UTSA L

WITHDRAWN
UTSA Libraries

International Political Economy Series

Series Editor: **Timothy M. Shaw**, Visiting Professor, University of Massachusetts Boston, USA, and Emeritus Professor, University of London, UK

The global political economy is in flux as a series of cumulative crises impacts its organization and governance. The IPE series has tracked its development in both analysis and structure over the last three decades. It has always had a concentration on the global South. Now the South increasingly challenges the North as the center of development, also reflected in a growing number of submissions and publications on indebted Eurozone economies in Southern Europe.

An indispensable resource for scholars and researchers, the series examines a variety of capitalisms and connections by focusing on emerging economies, companies and sectors, debates, and policies. It informs diverse policy communities as the established trans-Atlantic North declines and "the rest," especially the BRICS, rise.

Titles include:

Kobena T. Hanson, Cristina D'Alessandro and Francis Owusu (*editors*)
MANAGING AFRICA'S NATURAL RESOURCES
Capacities for Development

Daniel Daianu, Carlo D'Adda, Giorgio Basevi and Rajeesh Kumar (*editors*)
THE EUROZONE CRISIS AND THE FUTURE OF EUROPE
The Political Economy of Further Integration and Governance

Karen E. Young
THE POLITICAL ECONOMY OF ENERGY, FINANCE AND SECURITY IN THE UNITED ARAB EMIRATES
Between the Majilis and the Market

Monique Taylor
THE CHINESE STATE, OIL AND ENERGY SECURITY

Benedicte Bull, Fulvio Castellacci and Yuri Kasahara
BUSINESS GROUPS AND TRANSNATIONAL CAPITALISM IN CENTRAL AMERICA
Economic and Political Strategies

Leila Simona Talani
THE ARAB SPRING IN THE GLOBAL POLITICAL ECONOMY

Andreas Nölke (*editor*)
MULTINATIONAL CORPORATIONS FROM EMERGING MARKETS
State Capitalism 3.0

Roshen Hendrickson
PROMOTING U.S. INVESTMENT IN SUB-SAHARAN AFRICA

Bhumitra Chakma
SOUTH ASIA IN TRANSITION
Democracy, Political Economy and Security

Greig Charnock, Thomas Purcell and Ramon Ribera-Fumaz
THE LIMITS TO CAPITAL IN SPAIN
Crisis and Revolt in the European South

Felipe Amin Filomeno
MONSANTO AND INTELLECTUAL PROPERTY IN SOUTH AMERICA

Eirikur Bergmann
ICELAND AND THE INTERNATIONAL FINANCIAL CRISIS
Boom, Bust and Recovery

Yildiz Atasoy (*editor*)
GLOBAL ECONOMIC CRISIS AND THE POLITICS OF DIVERSITY

Gabriel Siles-Brügge
CONSTRUCTING EUROPEAN UNION TRADE POLICY
A Global Idea of Europe

Jewellord Singh and France Bourgouin (*editors*)
RESOURCE GOVERNANCE AND DEVELOPMENTAL STATES IN THE GLOBAL SOUTH
Critical International Political Economy Perspectives

Tan Tai Yong and Md Mizanur Rahman (*editors*)
DIASPORA ENGAGEMENT AND DEVELOPMENT IN SOUTH ASIA

Leila Simona Talani, Alexander Clarkson and Ramon Pachedo Pardo (*editors*)
DIRTY CITIES
Towards a Political Economy of the Underground in Global Cities

Matthew Louis Bishop
THE POLITICAL ECONOMY OF CARIBBEAN DEVELOPMENT

Xiaoming Huang (*editor*)
MODERN ECONOMIC DEVELOPMENT IN JAPAN AND CHINA
Developmentalism, Capitalism and the World Economic System

Bonnie K. Campbell (*editor*)
MODES OF GOVERNANCE AND REVENUE FLOWS IN AFRICAN MINING

Gopinath Pillai (*editor*)
THE POLITICAL ECONOMY OF SOUTH ASIAN DIASPORA
Patterns of Socio-Economic Influence

Rachel K. Brickner (*editor*)
MIGRATION, GLOBALIZATION AND THE STATE

Juanita Elias and Samanthi Gunawardana (*editors*)
THE GLOBAL POLITICAL ECONOMY OF THE HOUSEHOLD IN ASIA

Tony Heron
PATHWAYS FROM PREFERENTIAL TRADE
The Politics of Trade Adjustment in Africa, the Caribbean and Pacific

David J. Hornsby
RISK REGULATION, SCIENCE AND INTERESTS IN TRANSATLANTIC TRADE CONFLICTS

Yang Jiang
CHINA'S POLICYMAKING FOR REGIONAL ECONOMIC COOPERATION

Martin Geiger, Antoine Pécoud (*editors*)
DISCIPLINING THE TRANSNATIONAL MOBILITY OF PEOPLE

Michael Breen
THE POLITICS OF IMF LENDING

Laura Carsten Mahrenbach
THE TRADE POLICY OF EMERGING POWERS
Strategic Choices of Brazil and India

Vassilis K. Fouskas and Constantine Dimoulas
GREECE, FINANCIALIZATION AND THE EU
The Political Economy of Debt and Destruction

Hany Besada and Shannon Kindornay (*editors*)
MULTILATERAL DEVELOPMENT COOPERATION IN A CHANGING GLOBAL ORDER

Caroline Kuzemko
THE ENERGY–SECURITY CLIMATE NEXUS
Institutional Change in Britain and Beyond

Hans Löfgren and Owain David Williams (*editors*)
THE NEW POLITICAL ECONOMY OF PHARMACEUTICALS
Production, Innovation and TRIPS in the Global South

Timothy Cadman (*editor*)
CLIMATE CHANGE AND GLOBAL POLICY REGIMES
Towards Institutional Legitimacy

International Political Economy Series
Series Standing Order ISBN 978– 0–333–71708–0 hardcover
Series Standing Order ISBN 978– 0–333–71110–1 paperback
(*outside North America only*)

You can receive future titles in this series as they are published by placing a standing order. Please contact your bookseller or, in case of difficulty, write to us at the address below with your name and address, the title of the series and one of the ISBNs quoted above.

Customer Services Department, Macmillan Distribution Ltd, Houndmills, Basingstoke, Hampshire RG21 6XS, England

Managing Africa's Natural Resources

Capacities for Development

Edited by

Kobena T. Hanson
Head of Knowledge and Learning, African Capacity Building Foundation (ACBF), Zimbabwe

Cristina D'Alessandro
Knowledge Management Expert, African Capacity Building Foundation (ACBF), Zimbabwe

Francis Owusu
Professor, Iowa State University, USA

Editorial matter, selection, introduction, and conclusion © Kobena T. Hanson, Cristina D'Alessandro, and Francis Owusu 2014
Individual chapters © Respective authors 2014

All rights reserved. No reproduction, copy or transmission of this publication may be made without written permission.

No portion of this publication may be reproduced, copied or transmitted save with written permission or in accordance with the provisions of the Copyright, Designs and Patents Act 1988, or under the terms of any licence permitting limited copying issued by the Copyright Licensing Agency, Saffron House, 6–10 Kirby Street, London EC1N 8TS.

Any person who does any unauthorized act in relation to this publication may be liable to criminal prosecution and civil claims for damages.

The authors have asserted their rights to be identified as the authors of this work in accordance with the Copyright, Designs and Patents Act 1988.

First published 2014 by
PALGRAVE MACMILLAN

Palgrave Macmillan in the UK is an imprint of Macmillan Publishers Limited, registered in England, company number 785998, of Houndmills, Basingstoke, Hampshire RG21 6XS.

Palgrave Macmillan in the US is a division of St Martin's Press LLC, 175 Fifth Avenue, New York, NY 10010.

Palgrave Macmillan is the global academic imprint of the above companies and has companies and representatives throughout the world.

Palgrave® and Macmillan® are registered trademarks in the United States, the United Kingdom, Europe and other countries.

ISBN 978–1–137–36560–6

This book is printed on paper suitable for recycling and made from fully managed and sustained forest sources. Logging, pulping and manufacturing processes are expected to conform to the environmental regulations of the country of origin.

A catalogue record for this book is available from the British Library.

A catalog record for this book is available from the Library of Congress.

Library
University of Texas
at San Antonio

Contents

List of Tables, Figures, and Maps   vii

Acknowledgments   viii

Notes on Contributors   ix

List of Acronyms   xiii

1 Toward a Coordinated Approach to Natural Resource
   Management in Africa   1
   Kobena T. Hanson, Francis Owusu, and Cristina D'Alessandro

2 The Status of Natural Resource Management in Africa:
   Capacity Development Challenges and Opportunities   15
   Joseph Ayee

3 Governance of Natural Resource Management in Africa:
   Contemporary Perspectives   39
   Peter Arthur

4 Criminality in the Natural Resource Management
   Value/Supply Chain   66
   Olawale Ismail and Jide Martyns Okeke

5 Structured Transformation and Natural Resources
   Management in Africa   91
   William G. Moseley

6 Strategic Capacity-Building Imperatives Vital for
   Transboundary Water Cooperation in Africa   118
   Claudious Chikozho

7 The Gas and Oil Sector in Ghana: The Role of Civil Society
   and the Capacity Needs for Effective Environmental
   Governance   140
   Cristina D'Alessandro, Kobena T. Hanson, and Francis Owusu

8 The Capacity Question, Leadership, and Strategic Choices: Environmental Sustainability and Natural Resources Management in Africa   162
   Korbla P. Puplampu

9 Debating Critical Issues of Green Growth and Energy in Africa: Thinking beyond Our Lifetimes   185
   Abbi M. Kedir

10 Moving Africa beyond the Resource Curse: Defining the "Good-Fit" Approach Imperative in Natural Resource Management and Identifying the Capacity Needs   206
   Francis Owusu, Cristina D'Alessandro, and Kobena T. Hanson

Afterword: Natural Resource Governance Post-2015: What Implications for Analysis and Policy?   226
Timothy M. Shaw

Index   237

Tables, Figures, and Maps

Tables

8.1   Longevity of selected African rulers and the 2012 human
development status of their countries   168

Figures

5.1   Share of commodities in total exports by
developing region   94

5.2   Global urbanization patterns   95

5.3   Global commodity index adjusted for inflation   96

7.1   Risk cycle diagram   157

Maps

6.1   Main transboundary river basins in Africa   120

Acknowledgments

This book could not have been produced without the invaluable contribution of many individuals. To this end, we would like to express our sincere gratitude to all who variously and collectively helped make this publication possible. Throughout its development, we have benefited a great deal from the insightful commentaries and criticisms of colleagues who were generous enough to find time not only to read draft chapters and our initial thoughts, but also to offer us direction.

We are particularly indebted to the contributors. Their insightful chapters make a compelling argument for a rethink of natural resources landscape in contemporary Africa, and the development of an alternative framework to assess natural resources governance. We would also like to express our gratitude to the anonymous reviewers whose invaluable comments helped shape the quality and critical thoughts of the volume. Our appreciation further goes to our respective institutions: the African Capacity Building Foundation and Iowa State University for providing the enabling environment that allowed us to successfully collaborate on this initiative through fruition.

Heartfelt thanks go to Timothy M. Shaw, Series Editor, and to Christina Brian and Amanda McGrath at Palgrave Macmillan, for their support, encouragement, and guidance throughout the development of this volume.

To our families and loved ones, who sustained and encouraged us to keep chipping away, you have been the "wind beneath our wings." Thank you!

The views expressed remain the exclusive responsibility of the individual authors.

Notes on Contributors

Peter Arthur is Associate Professor of Political Science and International Development Studies at Dalhousie University, Canada. He received his PhD in Political Science in 2001 from Queen's University, Canada. His research interests focus on sub-Saharan Africa, with emphasis on the contribution of small-scale enterprises, electoral politics, capacity development and post-conflict reconstruction, the governance of oil, and natural resources management. His work has been published in a number of edited volumes and journals, including *Africa Today*, *African Studies Review*, *Commonwealth and Comparative Politics*, and *Journal of Contemporary African Studies*.

Joseph Ayee is Professor/Rector, Mount Crest University College, Accra, Ghana, and the first Emeka Anyaoku Visiting Professor of Commonwealth Studies, University of London. He is the immediate past Deputy Vice Chancellor and Head, College of Humanities, University of KwaZulu-Natal, South Africa. He also served as Head, Department of Political Science, and Dean, Faculty of Social Sciences, University of Ghana, Legon. He has published extensively in the fields of politics, governance, and development management in Ghana. His current articles have appeared in the *International Journal of Public Administration*, *The Round Table: The Commonwealth Journal of International Affairs*, and the *Journal of Asian and African Studies*. His research interests include political economy of mining, leadership, and the developmental state in Africa.

Claudious Chikozho is Director of the Exxaro Business and Biodiversity Leadership Program at the Centre for Responsible Leadership, University of Pretoria. He holds a PhD in Applied Social Sciences, a master's in Public Administration, and a BSc in Political Science from the University of Zimbabwe. His previous positions include working as a Research Fellow in natural resources management and Lecturer in development studies at the University of Zimbabwe; Senior Researcher in Water Governance at the Council for Scientific & Industrial Research in South Africa; and Science Uptake Coordinator for Africa at the International Water Management Institute in Accra, Ghana. His current areas of special interest include public sector leadership and service delivery, environmental sustainability, adaptation to climate change, and technology transfer processes.

Cristina D'Alessandro is a Knowledge Management Expert at the African Capacity Building Foundation (ACBF), Harare, Zimbabwe. She holds a PhD in Geography from the Université François Rabelais de Tours and a postdoctorate degree from West Virginia University (USA). She has published a number of critical papers and books. Her research is focused on the African continent and more precisely on political issues (at the local, national, as well as international level) and on institutional capacity building. She works as an external consultant for the African Capacity Building Foundation since 2011.

Kobena T. Hanson is Head, Knowledge and Learning Department, African Capacity Building Foundation (ACBF), Harare, Zimbabwe. He holds a PhD from Queen's University in Kingston, Ontario, Canada. He has published a number of critical articles on livelihood strategies and social networking in top-tier development policy and geography journals. His most recent book—*Rethinking Development Challenges for Public Policy* (coedited with George Kararach and Timothy M. Shaw)—is published by Palgrave Macmillan.

Olawale Ismail is currently the Head of Research at International Alert, London, UK. He is also an Associate Researcher with the Conflict, Security and Development Group (CSDG), King's College London, and the Stockholm International Peace Research Institute (SIPRI). Before now, he worked in a research capacity at the Stockholm International Peace Research Institute (SIPRI), CSDG, KCL, and the New York-based Social Science Research Council. His research interests include radicalization and counterterrorism, regional security mechanisms, peace-building and post-conflict reconstruction, peace operations, security sector and justice reform, disarmament, demobilization and reintegration, military expenditure and youth, and political violence.

Abbi M. Kedir is an associate member of staff at the School of Management, University of Leicester, UK. In the same university but in the Department of Economics, he served as an assistant professor from October 2003 to February 2013. He holds an MSc in Economics and Econometrics (1998) and a PhD in Economics (2003), both from the University of Nottingham, UK. He is also a freelance consultant. He is an applied development economist and works on policy-relevant analytical topics such as poverty, growth, saving, health, quadratic Engel curves, food security, aid coordination, foreign direct investment, trade, regional integration, and employment. Africa is his geographical focus and most of his published work is on Ethiopia—the country of his expertise. Among others, he has published in *Oxford Bulletin of Economics and Statistics*, *Journal of African Economies*, *Journal of Development Studies*, *European Journal of Industrial Relations*, *International Planning Studies and African Development Review*, and *Ethiopian Journal of Economics*. Since 1999, he has been a consultant to various institutions such as the World Bank, DFID, EU, African Capacity Building Foundation, African Economic Research Consortium, and African Development Bank.

William G. Moseley is Professor and Chair of Geography, and Director of African Studies, at Macalester College in St. Paul, MN, USA. He is an environment and development geographer with particular expertise in political ecology, tropical agriculture, food security, land reform, and Africa. Before becoming an academic, he worked for ten years in the field of international development for such organizations as Save the Children (UK), the World Bank, and USAID. He has published seven books and over 70 peer-reviewed articles and book chapters. He formerly served as a national councilor to the Association of American Geographers and editor of the *African Geographical Review*, and coauthored a National Academy of Sciences report on strategic directions for geography. His research has been funded by the National Science Foundation and the Fulbright-Hays program.

Jide Martyns Okeke is a Senior Civilian Planner for Peace Support Operations at the African Union Commission, Addis Ababa, Ethiopia. Concurrently, he is a Research Fellow at the Department of International Relations, University of Witwatersrand, South Africa. Previously, he was a Senior Researcher on African Peace and Security Architecture at the Institute for Security Studies (ISS), Addis Ababa office. He holds a PhD in Politics and International Studies from Leeds University and MA in Conflict Resolution (Bradford University), both from the UK.

Francis Owusu is Professor and Chair, Department of Community and Regional Planning, Iowa State University. He was an Assistant Professor of Geography and Political Science at Seattle Pacific University and Research Analyst at the Environmental Quality Board of Minnesota Planning, St. Paul, Minnesota. He has a PhD in Geography from the University of Minnesota. His research focus includes development policy, public sector reforms and capacity building, and urban development and livelihood issues. He has researched on these topics in several African countries and authored several journal articles, book chapters, book reviews, and reports. He has consulted several International Development Agencies, including the World Bank and African Capacity Building Foundation. He chaired the Africa Specialty Group of the Association of American Geographers (2005–2008).

Korbla P. Puplampu is in the Department of Sociology at Grant MacEwan University, Edmonton, Canada. He has a PhD in Sociology as well as an MEd in International and Global Education. His research interests are in the politics of knowledge production and propagation, global citizenship and identity politics in multicultural societies, as well as the global restructuring of higher education and agriculture. Articles have appeared in *Canadian Journal of Learning and Technology* and *Alberta Journal of Educational Research*, and he has coedited and contributed to several books including *The Public Sphere and the Politics of Survival: Voice, Sustainability and Public Policy in Ghana* (with Wisdom Tettey) and *African Education and Globalization* (with Ali Abdi and George Dei).

Timothy M. Shaw is Graduate Program Director of a new PhD program in Global Governance and Human Security at the University of Massachusetts Boston, USA. He is also visiting professor at Aalborg University, Denmark; Carleton University, Canada; Mbarara University, Uganda; and Stellenbosch University, South Africa. He has published in *Contemporary Politics, Journal of China and International Relations, Journal of Chinese Political Science,* and *Third World Quarterly*. Tim is Emeritus Professor at the University of London, UK, and holds an honorary degree from St Andrew's University, Scotland, UK.

Acronyms

ACBF	African Capacity Building Foundation
ACIR	Africa Capacity Indicators Report
ACSP	African Carbon Support Programme
AfDB	African Development Bank
AfGF	AfDB's African Green Fund
ALSF	African Legal Support Facility
AMDC	African Minerals Development Centre
AMV	Africa Mining Vision
APRM	African Peer Review Mechanism
AUC	African Union Commission
BRICS	Brazil, Russia, India, China, and South Africa
CADSP	Common African Defence and Security Policy (African Union)
CAR	Central African Republic
CCM	Chama Cha Mapunduzi
CD4CDM	Capacity Development for the Clean Development Mechanism
CDKN	Climate and Development Knowledge Network
CDM	Clean Development Mechanism
CDSF	Capacity Development Strategic Framework
CERs	Certified Emission Reduction units
CIFs	Climate Investment Funds
CoP	Conference of the Parties
CSOs	civil society organizations
CSPOG	Civil Society Platform on Oil and Gas
CSR	corporate social responsibility
DCED	Donor Committee for Enterprise Development
ECA	Economic Commission for Africa
ECOWAS	Economic Community of Western African States
EIA	environmental impact assessment
EITI	Extractive Industries Transparency Initiative
EU	European Union
FAO	Food and Agriculture Organization
FDI	foreign direct investment
GEF	Global Environmental Facility

GEMAP	Governance and Economic Management Assistant Programme (Liberia)
GHG	greenhouse gas
GNPC	Ghana National Petroleum Corporation
GPF	World Bank's Governance Partnership Facility
GWVSA	Global West Vessel Specialist Agency
IMF	International Monetary Fund
KPCS	Kimberley Process Certification Scheme
LCBC	Lake Chad Basin Commission
MDGs	Millennium Development Goals
MNCs	multinational corporations
NBI	Nile Basin Initiative
NEITI	Nigerian Extractive Industries Transparency Initiative

1
Toward a Coordinated Approach to Natural Resource Management in Africa

Kobena T. Hanson, Francis Owusu, and Cristina D'Alessandro

Introduction

Natural resource endowments can spur development in African countries, if managed appropriately. That said, resource exploitation does not automatically translate into meaningful development. Indeed, poor growth rates, high inequality, social exclusion, impoverishment, poor governance, environmental concerns, social tensions, and civil strife characterize many resource-rich countries across Africa and in the developing world (Collier 2007). These challenges have led to predictions of "natural resource curse" or "paradox of plenty" (Humphreys et al. 2007; Barma et al. 2012). Yet, while challenges to the effective management of natural resources—the so-called resource curse—are widely known, there exist critiques, theories, and models, some of which are subsets to, or overlap with, the resource curse while others are separate theories altogether (Humphreys et al. 2007; Obi 2010; Arthur 2012) that question the validity and accuracy of resource curse theorists. Humphreys et al. (2007) question the resource curse argument, because for them there is considerable room for human agency to correct the risks posed by the "paradox of plenty." For Africa more specifically, it has been demonstrated that the resource curse paradigm hides the larger question of how institutions and their transformation affect growth (Jones 2008). Other authors also insist that contextual variables encompassing the national level must be taken into account in explaining why natural resources are sometimes detrimental to the development of a country (Basedau 2005).

Such criticisms against the resource curse theorists have led to a shift away from the initial debates over the "greed versus grievance" binary. Hence, today, focus is mainly on issues related to capacities (individual, institutional, and enabling environment), leadership, and governance as the key drivers to negotiate the previously well-documented challenges of the natural resource sector (African Capacity Building Foundation (ACBF) 2013). As Barma et al. (2012: 4) argue, negotiating the so-called curse is "inherently a governance challenge: the credibility, quality, transparency, and accountability of policy-making processes, public institutions, the legal and regulatory climate, and sector governance are major determinants of how successfully countries can channel their resource wealth into sustainable development."

Toward a coordinated approach to resource management

In today's post-bipolar era, the mix of fragile/failed states (ACBF 2011), proliferating "global" issues, and pressures for democratization has generated some innovative forms of "transnational" (Brown 2011) or "private" (Dingwerth 2008) governance systems, symbolized by the Ottawa and Kimberley Processes, now augmented by the Programme for the Endorsement of Forest Certification (PEFC)/United Nations Collaborative Programme on Reducing (REDD) and Extractive Industries Transparency Initiative (EITI). Together with other anticorruption efforts and environmental concerns, these developments are changing the natural resource governance landscape (Shaw and Fanta 2013). They have served to encourage interstate international law toward the recognition of varieties of international governance, which may reflect varieties of sources of pressures. The need for Africa to advance capacity for strategic thinking and transformative leadership as the EU economies, and more broadly the Organisation for Economic Co-operation and Development (OECD), seek to reinvent themselves is apparent (Hanson et al. 2012). This capacity need is also justified by the emerging new BRICS model, by reshaping global governance, and by a landscape of more frequent crises and increased economic competition, in which African countries have to position themselves.

Africa can benefit from a strategy that, among others, depoliticizes the use of natural resource revenues by vesting their disposal in a sovereign authority set up along the lines of an independent yet accountable central bank, judiciary, and supervisory authority (Gylfason 2011). For this to happen and function properly, this volume emphasizes the need for a coordinated approach to natural resource management (NRM) in Africa.

The volume juxtaposes a set of "dynamic" relations—civil societies, value chains, green growth, leadership and sustainability, criminality, transboundary resource management, environmental governance, and structural transformation—with fluid interpretations of natural resource governance. The capacity-leadership-governance trichotomy represents the "axis of transformation" for the natural resource sector, especially the extractives, across resource-rich African states.

Drawing on the work of Alao (2007: 16), the volume conceptualizes natural resources to include "all non-artificial products situated on or beneath the soil, which can be extracted, harvested, or used, and whose extraction, harvest, or usage generates income or serves other functional purposes in benefiting mankind." We also view NRM from a technocratic perspective, and regard natural resource governance as a more comprehensive political process, involving myriad stakeholders. Accordingly, natural resource governance encompasses the framework of rules, institutions, and practices regulating the natural resource value chain and the extent to which key principles of transparency, openness, accountability, fairness, and environmental sustainability are observed in the extraction of, movement of, and receipts from natural resources. Environmental governance, employed in this volume specifically to the "management" of oil and gas industry (Chapter 7) with respect to the environment, rests on three key pillars: transparency and economic responsibility, environmental sustainability, and responsible community development (World Bank 2010).

Capacity, on the other hand, refers to

> the ability of people, organizations, and society as a whole to manage their affairs successfully; and that is the process by which people, organizations, and society as a whole unleash, strengthen, create, adapt, and maintain capacity over time. Capacity for individuals, organizations, and societies to set goals and achieve them; to budget resources and use them for agreed purposes; and to manage the complex processes and interactions that typify a working political and economic system. Capacity is most tangibly and effectively developed in the context of specific development objectives such as delivering services to poor people; instituting education, public service, and health care reform; improving the investment climate for small and medium enterprises; empowering local communities to

better participate in public decision making processes; and promoting peace and resolving conflict.

(ACBF 2011: 30–31; ACBF 2013)

Despite the multiple meanings of leadership, in developing contexts, it "is the process of organizing or mobilizing people and resources in pursuit of particular ends or goals, in given institutional contexts of authority, legitimacy and power (often of a hybrid kind)" (Lyne de Ver 2009: 9). Kaufmann, Kraay, and Mastruzzi (2008: 7) define governance as

> the traditions and institutions by which authority in a country is exercised. This includes the process by which governments are selected, monitored and replaced; the capacity of the government to effectively formulate and implement sound policies; and the respect of citizens and the state for the institutions that govern economic and social interactions among them.

If the transformation going through capacity, leadership, and governance is still a challenge for resource-rich African countries, affecting individuals, institutions, and organizations, some positive changes are already taking place across the African continent, and natural resources certainly play a role in these dynamics.

New dispensation in NRM in Africa

A wave of new optimism is sweeping across Africa—gross domestic product (GDP) is rising, consumer spending is increasing, and returns on investments are higher than global averages (McKinsey Global Institute 2010). Recent publications such as the *2013 Africa Capacity Indicators Report* (ACBF 2013), *Economic Report on Africa* (ECA/AU 2013), *Africa Progress Report 2013* (Africa Progress Panel 2013), and the *2014 Resource Governance Index* (RWI 2013) all highlight the evolving natural resource landscape in Africa. RWI (2013), for instance, notes that Ghana, Guinea, Liberia, South Sudan, and Zambia have all recently reformed their oil or mining legislation to include some principles of open government. Similarly, the growing exposure of the problem of "missing revenues" and growth of corporate social responsibility programs, extensive and participatory discussions of value chains and jobs, development of trust funds/sovereign wealth funds (SWFs), the increasing mobilization of community interests, and community-based NRM are all images of this evolving landscape (ACBF 2013). Another manifestation of the changing landscape is the Africa Mining Vision (AMV), which was adopted by African Heads of States and Governments in February 2009 (ECA/AU 2013). The AMV is perceived as a "driver for a fundamental and structural transformation of African economies, based on establishing and harnessing linkages between different economic sectors and regions" (Africa-Canada Forum 2013: 3).

Even as initiatives such as the EITI, Publish What You Pay (PWYP), and the Kimberley Process Certification Scheme (KPCS) continue to monitor resource extraction activities, the United Nations Environment Programme (UNEP) International Resource Panel (IRP) is steadily building up an understanding of global resource flows and why economic growth needs to be decoupled from rates of resource extraction (Swilling 2012). However, for the aforementioned initiatives, and others such as the Natural Resource Charter (NRC) (www.naturalresourcecharter.org) and AMV to really benefit Africans as a whole, there is the "need for greater ownership and buy-in by African citizens…and greater policy space for [countries] to regulate and monitor resource extraction for the benefit of their populations" (Africa-Canada Forum 2013: 10).

The negative situation of resource-rich economies in Africa is not immutable (Obi 2010; Arthur 2012; ACBF 2013). Many agree that better governance, transparency, and accountability are central to good resource management (NRC 2010), and can help enhance the potential value of natural resource endowments (Collier 2010). Studies suggest that the legacy of the "old" NRM landscape—characterized by asymmetries of weak states versus strong external actors multinational corporations (MNCs), consumer countries); low and often erratic commodity prices; unfair terms of trade; low technological and managerial capacity; insufficient and ineffective legal frameworks and policies; weak bargaining capacity and systems of taxation; lack of transparency and accountability across the value chain; windfall rents, when realized, extracted for the benefit of only the elite; lack of economic diversification and shared growth; unmitigated environmental damage caused by the extraction process; and the sociocultural displacement of affected communities—is giving way to an evolving positive landscape (Arthur 2012; Barma et al. 2012; ACBF 2013).

Today, internal and external pressures have resulted in a growing state coherence with strong policy frameworks and increased regional and subregional integration and linkages. This, coupled with high, if still erratic, commodity prices on Africa's extractives—driven by a strong demand from BRICS; advancements in technical skills (law, science, management); a recognition of need for training programs; extensive discussion of value chains and jobs; development of trust funds/SWFs; the promotion of green growth, REDD+, and national and international nongovernmental organizations' (NGOs') environmental advocacy; a proliferation of civil society organizations (CSOs) with increased capacity and international linkages; and new configurations of dialogue (among public sector, private sector, NGOs, and local communities)— have contributed to the increased rejection of blood diamonds, conflict timber, and abuses of MNC oil and gas extraction, even as national governments advance initiatives such as the African Peer Review Mechanism (APRM), AMV, African Minerals Development Centre (AMDC), and the African Development Bank's (AfDB) African Legal Support Facility (ACBF 2013). Each of these governance processes includes African regional programs to which ACBF relates with other African networks, including diasporas.

The entire portfolio of stakeholders is subject to a new normative environment, as evidenced by the strong NRM policy environment emerging across Africa (RWI 2013). Policies frame the sphere of potential constructive action, as this volume's chapters note; there are now new spaces for agency. The ongoing increases in the global prices of extractives, coupled with the expansion of new discoveries, and a growing demand from emerging economies represent an unparalleled opportunity for Africa's resource-rich states to bolster transformation within the ongoing evolving landscape. Enhancing the investment climate is one of the central steps to advancing international competitiveness in Africa's resource-rich economies (Page 2008). Countries need to advance and entrench policies that acknowledge the realities of their national contexts, that can bring about rapid results in a context of urgent need, and that allow for incremental improvements to their governance processes (Marcel 2013: 2).

So, while the ongoing developments are in the right direction, resource-rich African states need to embrace policies and initiatives that aim to sustain the current momentum. As Page (2008: 2) points out, much of Africa's post-1995 growth acceleration has been primarily driven by "avoiding the policy mistakes that led sharp economic contractions in the past and by a strong surge in growth in the resource-rich economies" (see also Arbache and Page 2007). To sustain and enhance the gains already achieved, resource-rich African states must also seek creative ways to overcome the knowledge asymmetries that they face in negotiations with foreign stakeholders, even as they develop appropriate tax structures, and invest in advancing capacity in all sectors along the entire natural resources value chain (Hanson and Léautier 2011; ACBF 2013; ECA/AU 2013; Marcel 2013; RWI 2013).

As Marcel (2013: 9) further argues, "instead of encouraging [African countries] to pursue 'best practice' standards, it may be more helpful to advise them to aim for *'more appropriate practice'*, which acknowledges the realities of the national context; *'more effective practice'*, which seeks to bring about rapid results in the context or urgent needs, or *'better practice'*, which aims at incremental improvements of governance processes through aspirational, but achievable, milestones." Marcel, thus, concurs with Barma et al. (2012), and this volume's editors, who all call for a "good-fit" approach to NRM (see Chapter 10 by Owusu, D'Alessandro, and Hanson, in this volume).

Viewpoints and reflections

The need for Africa to develop critical capacity at the individual, institutional, and enabling environment levels for NRM that ensures transformation is apparent. Mitigating the so-called African paradox of plenty is a leadership, governance, and capacity challenge: effective auditing, monitoring, regulating, and improving resource exploitation regimes and developing resource sector linkages into diversified domestic economy; advancing capacity of the legislature to act as a countervailing force over the executive; enhancing budgeting and expenditure management; improving procurement practices and grants of natural resources concessions; establishing effective mechanisms to curb corruption; and supporting central institutions of government (Barma et al. 2012; ACBF 2013). Africa thus needs to work toward advancing knowledge sharing and enhancing transformative leadership and institutions that further its development agenda and goal of good governance, reducing poverty and inequalities, and improving the quality of life of African populations through natural resources. It is in this light that this volume's collection of essays seeks to assist Africa claim the 21st century.

The volume synthesizes viewpoints and reflections by drawing on the extant literature on NRM and incorporating rich empirical material and case studies from a host of countries to reflect the African experience in its complexity and diversity. The chapters of this book provide snapshots of the several emerging "worlds" of NRM, and policy responses to a set of new global issues and coalitions. They identify salient capacity development strands for the myriad stakeholders involved in the NRM value chain. In turn, they highlight a set of relevant, revisionist, often overlapping analyses of both the governance challenge and the emerging revised landscape—drawing on authors' granular understanding of the African NRM landscape.

Chapters 2 and 3 both interrogate Africa's natural resource landscape from perspectives that highlight the evolution of a revised landscape, and question the resource curse theory. Joseph Ayee (Chapter 2) explicitly examines the NRM landscape in Africa in relation to the perceived resource curse hypothesis, and draws on case studies and the extant literature to argue that the natural resource curse is not inevitable. Rather, issues of capacity, leadership, and good governance are what need privileging. Specifically, Ayee examines the role the postcolonial state has played in Africa vis-à-vis MNCs in the exploitation of natural resources on the continent. The discussion is situated in the historical context with a view to answering the key questions: (i) Why is there a renaissance in NRM in Africa today? (ii) How does Africa's natural resource sector compete with its international comparators? (iii) What are the capacity development challenges and opportunities on the continent as they relate to NRM? The chapter cites poor governance as one of the weakest links in the natural resource sector in Africa, and puts forth an overarching framework of governance as a way forward.

In Chapter 3, Peter Arthur takes up in more detail the status of NRM in Africa, examining the capacity development challenges and opportunities. The chapter reviews the literature on the resource curse and proposes some necessary interventions that could enhance effective management of resources in African countries. Notable are the promotion of transparency and accountability, the rule of law, as well as central roles for media and civil society groups and initiatives (EITI, KPCS, PWYP), which can act as watchdogs over African countries' and foreign stakeholders' activities in the natural resource sector. The chapter also takes the position that, given the limited capacities of many African countries, capacity development measures are needed to ensure that the vast information and power asymmetries between African countries and their counterparts in other parts of the world are addressed.

Given the importance of efforts at the national level to inform regional and global initiatives, a number of authors, notably Ismail and Okeke (Chapter 4), and D'Alessandro, Hanson, and Owusu (Chapter 7), highlight that multiple scales must be taken into account concurrently. Ismail and Okeke's exposition of criminality in the natural resource value chain in Africa, using timber (logging in Liberia) as an example of renewable, and oil (in Nigeria) as a nonrenewable natural resource, emphasizes that criminality in the natural resource value chain in Africa is multilayered, transcends conventional categorizations, and is hierarchically structured to include local, national, regional, and international actors. Ismail and Okeke further tease out the nature, scale, and dimensions of criminality and provide invaluable insights into the multi-scalar—local, national, regional, and international—nature of criminality in the sector. D'Alessandro, Hanson, and Owusu (Chapter 7) discuss the role of civil society organizations and explore their potential contributions to improving Ghana's nascent oil industry, especially in relations to social and environmental concerns. The chapter acknowledges that a great deal of capacity is equally required on the part of international investors, who often are unable/unwilling to grasp local sociopolitical, ethno-cultural, and economic environments so as to be able to innovate and derive mutually beneficial arrangements.

African economies, it has been argued, must diversify away from primary production (resource extraction and agricultural commodity production) if they are to improve their economic position vis-à-vis the rest of the world (ACBF 2013; ECA/AU 2013). Some African economies have begun to diversify by developing more extensive value chains in existing extractive industries, whereas others have seized on new windows of opportunity, notably the industry-supported Programme for the Endorsement of Forest Certification (PEFC) and the G8-supported EITI, the latter being particularly timely given the dangers of "resource curse" and windfall profits around Brazil, Russia, India, China, and South Africa (BRICS) demand for energy and minerals (Hanson et al. 2012). Elbadawi and Kaltani (2007), however, note that successful booming sector-driven economic transformation also demands macroeconomic and financial frameworks for promoting national savings, fiscal stability, diversification, and a political and social contract for managing booming sector revenues, based on democratic participation and transparent economic governance. Moseley (Chapter 5) takes up this issue in examining recent trends in African resource-based economies and the risks of an economy overly focused on primary production, and interrogates past approaches that have been undertaken to pursue economic diversification (failed and successful), to answer the question: Do natural resources intrinsically impede economic diversification?

Transnational governance of natural resources has emerged as a major driver of development in Africa and an effective way to implement regional integration across the continent. This notwithstanding, it has been argued that there is need for caution on the transnational governance framework Africa engages in (Mayntz 2010; Hanson et al. 2012). Democracy requires that there must be systems of accountability and transparency. Yet experience suggests that supranational institutions are not held accountable to voters, lack sanctioning power, and are often without institutionalized forms for the direct expression of popular preferences (Menon and Weatherill 2008). Claudious Chikozho (Chapter 6) attempts to navigate this dilemma in exploring the challenges and opportunities evident in transboundary water resources management and institutional capacity building. Chikozho submits that myriad administrative, managerial, technical, financial, and knowledge-related challenges act as barriers to effective transboundary water governance in Africa. The adoption of a more systematic approach to capacity building based on lessons learnt from successful transboundary river basins that have had a long history of interstate cooperation is proposed as a solution. Finally, the chapter cautions that the main thrust of the management of shared river basins should be to find ways of turning potential conflicts into constructive cooperation, in which riparian countries share the costs and benefits of water development projects.

Puplampu (Chapter 8) focuses on the role of political leadership and policy choices in managing natural resources by examining the relationship between capacity building at the individual, institutional, and broader environment levels and NRM in Africa. The chapter argues that political leadership sets the tone for the institutional and broader environment required for environmental sustainability and development. Puplampu concludes that in the final analysis, it is not only imperative to have policies and institutions, but also leaders of goodwill whose raison d'être is the national good. Kedir (Chapter 9) interrogates the emerging issues of green growth/green economy and how they relate and impact NRM in Africa. In so doing, he cautions that "growing dirty and cleaning up later" is no longer a viable option. Owusu, D'Alessandro, and Hanson (Chapter 10) conclude the volume, highlighting the capacity imperatives required to transcend the so-called resource curse. The chapter advocates for a political economy approach to NRM that is premised on the need for "good-fit" policies rather than "best practices."

Concluding remarks

Africa is at a crossroads in terms of growth, development, governance, and sustainability (Hanson et al. 2012): Can it seize its chances and transcend its somewhat lackluster first half-century? The authors of this volume, individually and collectively, believe so. Evidence put forth suggests that from Africa is emerging a new, more complex, evidence-based, participatory, and coordinated vision of NRM. This vision of natural resource-based development is motivated by an increasingly diversified and empowered portfolio of stakeholders and actors. The implication is that real growth and transformation based on natural resources is possible.

Numerous and important opportunities for turning Africa's natural resource wealth into sustainable and broad-based development exist. Africa can benefit from "new donors" and "innovative sources of finance" as it seeks to revise its natural resource sector: from SWFs and new foundations to the Gulf states, Korea, Turkey, as well as the BRICS. New technologies already allow for the discovery of new sources of oil and gas from South Sudan to northern Mozambique, so new regions may arise around pipelines and other energy logistics/corridors (Ulrike and Rempe 2013). Also, novel forms of "global" or "private" governance, from the Kimberley Process (KP) and the EITI to the "conflict minerals" component of the omnibus Dodd-Frank bill, while insightful, present challenges of compliance and reporting (Hanson et al. 2012). Accordingly, other contributions, in a forward-looking stance, call for rethinking and redoubling of efforts by Africans to advance leadership capacity, as well as individual and institutional capacity to tackle emerging issues and challenges, and more importantly pursue a "good-fit" rather than a "best practice" approach to the management and transformation of the continent's vast and diversified natural resource endowments.

High levels of poverty, rising inequalities, unsustainable unemployment rates (especially among the youth and vulnerable populations), corruption, and environmental and climate change concerns are contributing to eroding the quality of life and life expectancy of the vast majority of African people. African dynamics such as an increased demographic pressure on spaces where the resources are concentrated, and internal migrations that escape any accurate monitoring and control are becoming sensitive problems even far beyond the African continent.

Natural resources can and have to be propelling instruments, helping African countries to improve their economies in a sustainable way, to contribute to reform institutions to make them efficient and effective. Given their crucial commercial weight, they offer a chance to enhance a virtuous circle, in which better practices encourage a more engaged leadership to emerge, more accountability and transparency at the national level, more active, organized, and capable CSOs, and an equally more powerful private sector, helping to diversify the economies and fight successfully against widespread unemployment and dangerous social and spatial inequalities. This is possible because, as this volume argues, Africa is already on a good path: these efforts must be continued, but buttressed by knowledge sharing and cooperation.

References

ACBF (African Capacity Building Foundation). (2011). *Africa Capacity Indicators 2011: Capacity Development in Fragile States,* Harare: ACBF.

ACBF. (2013). *Africa Capacity Indicators 2013: Capacity Development for Natural Resource Management,* Harare: ACBF.

Africa-Canada Forum. (2013). *The African Mining Vision: A Transformative Agenda for Development,* Available at: http://www.ccic.ca/_files/en/working_groups/2013-04-02-AMV_backgrounder_EN.pdf. Accessed October 4, 2013.

Africa Progress Panel. (2013). *Africa Progress Report 2013. Equity in Extractives: Stewarding Africa's Natural Resources for All,* Geneva: Africa Progress Panel.

Alao, C. (2007). *Natural Resources and Conflict in Africa: The Tragedy of Endowment,* Rochester, NY: University of Rochester Press.

Arbache, J.S. and Page, J. (2007). "More Growth or Fewer Collapses? An Investigation of the Growth Challenges of Sub-Saharan African Countries," Policy Working Paper no. 4384, Washington, DC: World Bank.

Arthur, P. (2012). "Averting the Resource Curse in Ghana: Assessing the Options," in L. Swatuk and M. Schnurr (eds.), *Natural Resources and Social Conflict: Towards Critical Environmental Security,* London: Palgrave Macmillan, pp. 108–127.

Barma, N.H., Kaiser, K., Le, T.M., and Viñuela, L. (2012). *Rents to Riches: The Political Economy of Natural Resource-Led Development,* Washington, DC: the World Bank.

Basedau, M. (2005). "Context Matters—Rethinking the Resource Curse in Sub-Saharan Africa," DUI Working Papers no. 1, Hamburg: German Overseas Institute.

Brown, S. (ed.) (2011). *Transnational Transfers and Global Development,* London: Palgrave Macmillan.

Collier, P. (2007). *The Bottom Billion: Why the Poorest Countries Are Failing and What Can Be Done about It.* New York: Oxford University Press.

Collier, P. (2010). "The Political Economy of Natural Resources," *Social Research,* 77(4): 1105–1132.

Dingwerth, K. (2008). "Private Transnational Governance and the Developing World," *International Studies Quarterly,* 52(3): 607–634.

ECA/AU (Economic Commission on Africa/African Union Commission). (2013). *Economic Report on Africa 2013: Making the Most of Africa's Commodities: Industrializing for Growth, Jobs and Economic Transformation,* Addis Ababa, Ethiopia: ECA/AU.

Elbadawi, I. and Kaltani, L. (2007). "Strategies for Managing the Current Oil Boom," *Paper presented at the AERC Senior Policy Seminar IX Seminar on Managing*

Commodity Booms in Sub-Saharan Africa, Yaoundé, Cameroon, February 27–March 1.

Gylfason, T. (2011). "Natural Resource Endowments: A Mixed Blessing?" CESifo Working Paper 3353, Munich: CESifo.

Hanson, K., Kararach, G., and Shaw, T. (eds.) (2012). *Rethinking Development Challenges for Public Policy: Insights from Contemporary Africa*, London: Palgrave Macmillan for ACBF.

Hanson, K. and Léautier, F.A. (2011). "Enhancing Institutional Leadership in African Universities: Lessons from ACBF's Interventions," *World Journal of Entrepreneurship, Management and Sustainable Development*, 7(2–4): 386–417.

Humphreys, M., Sachs, J.D., and Stiglitz, J.E. (2007). *Escaping the Resource Curse*, New York: Columbia University Press.

Jones, S. (2008). "Sub-Saharan Africa and the Resource Curse," Working Paper no. 14, Copenhagen: Danish Institute for African Studies.

Kaufmann, D., Kraay, A., and Mastruzzi, M. (2008). "Governance Matters VII: Aggregate and Individual Governance Indicators 1996–2007," Policy Research Working Papers no. 4654, Washington, DC: the World Bank.

Lyne de Ver, H. (2009). "Conceptions of Leadership," Background Paper no. 4, Canberra, Australia: Development Leadership Program, AusAID.

Marcel, V. (2013). *Guidelines for Good Governance in Emerging Oil and Gas Producers* (September), London: Chatham House (The Royal Institute of International Affairs).

Mayntz, R. (2010). "Legitimacy and Compliance in Transnational Governance," Working Paper 10/5, Köln, Germany: Max Planck Institute for the Study of Societies (MPIfG).

McKinsey Global Institute. (2010). *Lions on the Move: The Progress and Potential of African Economies*, McKinsey Global Institute, Available at: www.mckinsey.com/mgi.

Menon, A. and Weatherill, S. (2008). "Transnational Legitimacy in a Globalizing World: How the European Union Rescues Its States," *West European Politics*, 31(3): 397–416.

NRC (The Natural Resource Charter). (2010). *Natural Resource Charter* (November), Available at: www.naturalresourcecharter.org. Accessed October 3, 2013.

Obi, C. (2010). "Oil Extraction, Dispossession, Resistance, and Conflict in Nigeria's Oil-Rich Niger Delta," *Canadian Journal of Development Studies*, 30(1/2): 219–236.

Page, J. (2008). "Rowing against the Current: The Diversification Challenge in Africa's Resource-Rich Economies," Global Economy & Development Working Paper 29, Washington, DC: Brookings Institution.

RWI (Revenue Watch Institute). (2013). *The 2013 Resource Governance Index. A Measure of Transparency and Accountability in the Oil, Gas and Mining Sector*, Available at: http://www.revenuewatch.org/sites/default/files/rgi_2013_Eng.pdf.

Shaw, T.M. and Fanta, E. (2013). "Introduction: Comparative Regionalisms for Development in the 21st Century: Insights from the Global South," in E. Fanta, T.M. Shaw, and V.T. Tang (eds.), *Comparative Regionalisms for Development in the 21st Century*, Farnham: Ashgate, pp. 1–18.

Swilling, M. (2012). "Beyond the Resource Curse: From Resource Wars to Sustainable Resource Management in Africa," *Paper presented at the Winelands Conference on Integrity and Governance*, Stellenbosch, South Africa (April).

Ulrike, L. and Rempe, M. (eds.) (2013). *Mapping Agency: Comparing Regionalisms in Africa*, Farnham: Ashgate for UNU-CRIS.

World Bank. (2010). *Environmental Governance in Oil Producing Developing Countries: Findings from a Survey of 32 Countries*, Washington, DC: World Bank.

2

The Status of Natural Resource Management in Africa: Capacity Development Challenges and Opportunities

Joseph Ayee

Introduction

Africa's 56 countries are endowed with both renewable (water, land, forest, fish) and nonrenewable or depletable (minerals, metals, oil) natural resources that may be discovered, remain undiscovered, or barely harnessed. The continent contains more than half the world's resources of cobalt, manganese, and gold and as well as significant supplies of platinum, uranium, and oil. An estimated $1 trillion of minerals, metals, and oil were extracted in 2008, with commodity exports accounting for 38 percent of the continent's gross domestic product (GDP) (Forstater et al. 2010).

Natural resources can contribute to economic growth, employment, and fiscal revenue. They are often a major source of national income but are also, if mismanaged or shared unfairly, a major cause of conflict and instability (Auty 2001a, b; Collier and Hoeffler 2002). Countries with weak institutions often struggle to handle the potentially destructive force of corruption and efforts by various actors to capture the wealth generated by natural resources. The governance of natural resources is especially important in the context of divided societies, because control over the benefits from local natural resources is often a chief motivator of ethnic or identity-based conflicts (Custers and Matthysen 2009; Lesourne and William 2009). Many resource-rich and resource-dependent African countries are characterized by disappointing growth rates, high inequality, widespread impoverishment, bad governance, and an increased risk of civil violence (Collier 2007; Dunning 2008; Mildner et al. 2011). The challenges posed by managing natural resources have been characterized as the "natural resource curse" or "paradox of plenty" (Auty 1993; Humphreys et al. 2007; UNDP 2011).

Against this backdrop, this chapter examines the status of natural resource management in Africa and the capacity development challenges and opportunities. Specifically, it examines the role the post-colonial state in Africa has played vis-à-vis multinational corporations (MNCs) in the exploitation of natural resources on the continent. The discussion is also situated in the historical context with a view to answering the following four questions: (i) Why is there a renaissance in natural resource management in Africa today? (ii) How does Africa's natural resources sector compete internationally? (iii) Which capacity development challenges and opportunities on the continent relate to the natural resource management? and (iv) What overarching theoretical framework or model can be used to interrogate natural resource management in Africa? To consider these questions, it is imperative to unpack the colonial legacy to enable us better understand the relationship between the postcolonial state, MNCs, and natural resources.

The colonial legacy

The Berlin Conference of 1884–1885 marked the beginning of a new era of economic and political relationships between Europe and Africa. African colonies supplied inexpensively produced agricultural commodities, such as rubber and cotton, and such minerals and metals as gold, diamonds, manganese, and copper to industries in Europe. Manufactured textiles, household goods, and farm implements sold to Africans at high profit completed the integrated economic system (Hodder-Williams 1984; Young 2000).

The colonial policies worked to handicap independent Africa's economic future (Amin 1972; Rodney 1972). The departure of the colonial administrations left postindependence African economies distorted and lacking integration; decisions, strategies, and even sovereignty were contingent on foreign markets, finance, and expertise and the influence of MNCs (Ake 1981; DeLancey 2001).

The political structures that developed were essentially alien and hastily imposed. They were intended primarily to control the territorial population, implement exploitation of natural resources, and maintain themselves and the European population. The long-term experience with the colonial state also shaped the nature of ideas bequeathed at independence, as leaders became authoritarian, self-interested, and corrupt (Wunsch and Olowu 1990; Gordon 2001: 60).

The postcolonial African state

Much ink has been devoted to the postcolonial state in Africa, which has been characterized as "prismatic" (Riggs 1964), "soft" (Myrdal 1968), "weak" (Jackson and Rosberg 1982), "overdeveloped" (Leys 1976), "pre-capitalist affectation" (Hyden 1983), "anti-development" (Dwivedi and Nef 1982), "predatory" (Fatton 1992), and "vampire" (Frimpong-Ansah 1992). African countries inherited the idea of the state as an "engine of growth" from the former colonial rulers (Jackson and Rosberg 1982; Ergas 1987; Young 1994). With the exception of Tanzania, Guinea, and Zambia, which adopted a "socialist path" to development, most of the other countries declared a commitment to a "mixed" economy but with the state controlling the commanding heights, which did not work because of the colonial legacy, patronage, and lack of diversification of the economy (Sklar 1975). This created an economic crisis in the late 1970s and the 1980s, and with the lessons of international experience from the success of market-friendly economies combined to force a redefinition of the role of the state (Tangri 1999; Herbst 2000).

Given the incapacity of the state to implement structural adjustment programs (SAPs), the World Bank and other donors moved in the 1980s toward a concern with improving state capacity, that is, governments' capacity to achieve their stated objectives, design and implement economic policies for growth, and provide good governance to their societies and markets (Brautigam 1996; Englebert 2000). This was to be achieved by "rolling back the state"—that is, restricting the role of the state while providing greater opportunity for market forces to assert themselves on the development process and liberalizing the economy to induce economic development (Hyden 1983; Jeffries 1993). The concern also involved building administrative capacity as an instrument of the development process rather than of a spoils system and the development of more efficient and, in a sense, more autonomous state machines (Adamolekun 1999).

Various panaceas were suggested, including administrative reform covering areas such as organizational development, manpower development, training, and the introduction of management techniques along the lines of the new public management school (Turner and Hulme 1997; Levy, 2004). The question about the role of the state is also conceptualized as "matching the role to capability," which involves basic government tasks to be performed by the state: "a foundation of law, a benign policy environment, including macro-economic stability, investing in people and infrastructure, protection of the vulnerable, and protection of the natural environment" (World Bank 1997: 4–8).

By the late 1990s, the need for a market-friendly economy had become widely accepted throughout Africa. This implies a reduced role for the state in economic management. The state provides an enabling environment for private sector economic activities by implementing appropriate economic policy reforms and providing the necessary legal and regulatory framework (Adamolekun 1999; Tangri 1999). The pursuit of this business-friendly environment also led to proliferation of MNCs in the natural resource sector.

The state, multinational corporations, and natural resources in Africa

The involvement of the state in natural resource management in Africa may be divided into three phases. The first phase was geared toward the state's involvement in natural resource management. This was largely ideologically driven by newly independent countries that stressed the need for self-determination and control of the national patrimony. After gaining independence from colonial rule, natural resource-rich states in Africa established state-owned enterprises (SOEs) to exploit natural resources (Killick 1978; Herbst 2000). Governments believed at the time that the state should become an entrepreneur or engage in what one might call "state capitalism" (Mafeje 1977) or a statist approach to development. However, the SOEs did not perform well under the burden of their multiple objectives, which included generating surpluses, attaining a healthy balance sheet, achieving internal management efficiency, and promoting such social welfare goals as being avenues for employment and selling goods below existing market prices (Hyden 1983; Tangri 1999).

The second phase involved the nationalization of private corporations engaged in natural resource exploitation. This often occurred where radical and often military governments came to power. Nationalization of private companies took place under the "*Ujamaa,*" or African socialism ideology, of Julius Nyerere's Chama Cha Mapinduzi (CCM) government in Tanzania (Tandon 1979). African states' nationalization interventions did not last for long and have not yielded the expected dividends. This is because the states lack the resources (for instance, capital, technology, and personnel) to revamp the natural resources sector and also compete with MNCs (Dunning and Lundan 2008).

The third phase is the current dominance of MNCs, with the states playing only regulatory roles. This is the dominant view of the Washington Consensus, which throughout the 1980s advocated minimal state intervention in the market through privatization, deregulation, and liberalization.

There has been a flurry of scholarly works on MNCs and their role in the natural resources sector in Africa (Drucker 1974; Sklar 1975; Apter and Goodman 1976; Moran 1978; Leonard 1980; Alden and Davies 2006; Wiig and Kolstad 2010; Ozoigbo and Chukuezi 2011). MNCs have had a long history in Africa because of their role in exploiting natural resources in colonial and postcolonial periods. They are pervasive to the extent that some scholars have referred to them as Africa's "new colonizers" (Dunning 1993a, b) while others see them as the "Janus face" of globalization (Eden and Lenway 2001).

Their role in the economy and natural resources exploitation has attracted considerable debate among scholars (Vernon 1977; Collier 2011). Opinions vary on the extent to which Africa has benefited from natural resources and MNCs: some view them as real partners in the development process (Dunning 2008; Wiig and Kolstad 2010), and others view their contribution in the natural resources sector as a myth and counterproductive (Drucker 1974; Ozoigbo and Chukuezi 2011).

An issue that has dominated the debate is the balance of power between national governments and MNCs, which often appear to have a firm and unfair advantage. As a result, MNCs have developed a love–hate relationship with the host economies (Eden and Lenway 2001). The host governments feel cheated while the MNCs also feel that they are doing the right thing, although they can be said to have been dishonest with the host country from the onset of contract negotiations (Robinson 1979).

Although MNCs' activities on Africa's natural resources sector date to the colonial and postcolonial periods, the economic globalization launched in the 1980s expanded their reach and gave them continued momentum. More recently, several factors have contributed to the influx of a new wave of MNCs in Africa, including Japanese and Chinese state-sponsored companies. They include the following:

1. the combination of abundant resources and vast market potential (Wiig and Kolstad 2010);
2. continued deepening of political and economic reforms in African countries created the best ever environment for MNCs to invest in Africa (Jeffries 1993; Alden and Davies 2006); and

3. the interest in new business opportunities in Africa by some western and Asian countries that originally had maintained economic distance from the continent (Forstater et al. 2010).

The discussion so far should not be construed as a wholesale acceptance of MNCs' leadership in resource exploitation in Africa, nor should the states' leadership be pushed to the back burner. The state has a role to play in natural resource exploitation under certain conditions. Three key reasons underlie this point:

- First, the state must have the capital, technology, and entrepreneurial resources necessary to compete favorably with MNCs. Unfortunately, these have become "scarce commodities" that most African countries lack, which necessitates the development of a liberal legal and investor-friendly environment to attract foreign MNCs (World Bank 2009).
- Second, governments of most natural resource-rich countries in Africa should be advocating resource nationalism. This refers to a situation where producer countries have moved to maximize revenue from natural resource production while altering the terms of investment for future output as part of increased government intervention in resource development. The shift entails two components: (i) limiting the operations of MNCs, and (ii) asserting greater national control over natural resource development (Stevens 2008; Ward 2009).
- Third, governments should resort to the use of the obsolescing bargaining model, whereby once a natural resource has been discovered and the investment sunk in development, relative bargaining power switches in favor of the host government, which then tries to increase its fiscal take by unilaterally changing the terms of the original contract (Ramamurti 2001).

The renaissance in natural resources management in Africa today

The current upsurge in interest in natural resource management and the proliferation of MNCs is attributable to a number of factors.

First, despite having an abundant natural resource endowment, Africa has experienced disappointing results in translating this natural wealth into broad economic development. The net development impact of natural resources has been modest. For instance, natural resource-related taxes, which consist mainly of royalties and corporate income taxes, generated just 32 percent of Africa's GDP growth from 2000 through 2010 (AfDB 2010). In some countries, revenue from renewable natural resources is substantial, for instance, from fisheries in Namibia and forestry in Cameroon (AfDB 2010; UNDP 2011).

Second, the literature has shown that the role of natural resources in the promotion of economic growth has become one of the core issues of development theory and practice. It is argued that developed countries such as the United States, Canada, Australia, and the Scandinavian countries became rich and technologically advanced through a judicious use of their natural resources wealth (Lederman and Maloney 2007; Dunning 2008). This positive view of natural resources is also shared in the 2003 African Convention on the Conservation of Nature and Natural Resources.

Third is the growing trend in Africa to improve and maximize tax collection and use it to contribute to improved governance. Consequently, it has become imperative to improve natural resources taxation as an essential revenue source for many African countries, which hitherto has not been exploited in Africa (Brautigam et al. 2008; Pritchard 2010).

Fourth is the continuing presence of the two major negative effects from natural resource wealth: the resource curse and the Dutch disease, which have bedeviled most natural resource-rich countries in the continent (Strauss 2000; Sachs and Warner 2001; Gylfason 2004; Mehlum et al. 2006). Some authors have even classified natural resources as one of the ten most robust variables with a significantly negative effect on growth in empirical studies (Sala-i-Martin 1997; Doppelhofer et al. 2000: Gylfason 2001). This is because economic activities related to nonrenewable natural resources, especially, are perhaps the most susceptible to looting because the (i) resources are geographically fixed and cannot relocate; (ii) resource extraction requires relatively low ongoing operational investment to maintain the productivity of the initial physical infrastructure; and (iii) products are usually exported, which creates many choke points for extortion, such as pipelines, roads, and ports (Gylfason and Zoega 2006; Synder 2006).

Fifth is the continued challenge of the social and environmental costs associated with the exploitation of natural resources with little welfare-enhancing return. Few of the developmental benefits expected to accompany the exploitation of minerals have materialized in the host communities in spite of the launching of corporate social responsibility (CSR) initiatives launched by some MNCs to mitigate problems such as displacement of indigenous communities, loss of livelihoods, and adulteration of local culture; conflicts and human rights abuses; diversion of watercourses; and loss of biodiversity due to environmental destruction (Salami 2001; Akabzaa et al. 2007). In addition, the communities tend to expect the companies to provide basic amenities and infrastructure that the central government has failed to provide, as if they were surrogate governments (Akabzaa et al. 2007; Campell 2009).

This expectation led to the strike action by mine workers of Lonmin Platinum Company in Marikana, North West Province, of South Africa in mid-August 2012 and the consequent death of 45 people, including two policemen—the highest since the transition to majority rule in 1994. The mine workers demanded wage increase from R6,000 to R12,500 per month while Lonmin also failed to meet its social, economic, and environmental commitments within the framework of global CSR. In other words, social ills apart from the wage demand fueled the unrest:

> Marikana township constructed by Lonmin did not have electricity for more than one month, and at a nearby RDP township broken drains were spilling into river . . . In Marikana there are broken sewage systems, bilharzia in the water, children are getting sick . . . lack of educational facilities and training, environmental pollution . . . Many mine workers rented shacks in informal settlements and live in appalling conditions. All these led to tension in the communities and high youth discontent.
>
> (Macleod 2012: 2–3)

This incident fueled a spate of natural resources strikes, renewed calls in South Africa for transformation and nationalization of the mining industry, and exposed the industry's inability to live by its CSR as well as the government's incapacity to enforce mining law and regulation (Ramaphosa 2012). The one lesson that can be learned from the Marikana crisis is that expectations of returns from the natural resources sector have been particularly high at the community level in all African countries and have caused frustration among the youth in the communities and tension or rivalries among the various mining communities (Babu 2000; Ikelegbe 2006; Mildner et al. 2011).

A key governance issue is the historical bias of central government institutions in favor of corporate and especially transnational investors vis-à-vis the interest of the communities. The fact of the matter is that in Africa policy-making is centralized in national institutions that seem to have no direct accountability to communities or even the local government units (Ross 1999; Ariweriokuma 2009; Mildner et al. 2011).

Sixth is the proliferation of both local and international drivers of good natural resource management that, through their activities or enforcement, have put the matter on the agenda. These drivers may be roughly classified into nine groups: (i) international financial institutions, (ii) other multilateral initiatives, (iii) regional initiatives, (iv) bilateral donors, (v) nongovernmental organizations, (vi) industry groups, (vii) multi-stakeholder initiatives, (viii) charters and conventions, and (ix) in-country civil society organizations. The drivers have no doubt shifted "attitudes in resource-rich developing countries...where, for instance, the EITI served as a concrete rallying point for both reformist countries and for reformers in reluctant countries" (UNECA 2009: 230–231) while "an international charter gives people something very concrete to demand: either the government adopts it or it must explain why it won't" (Mildner et al. 2011: 164–165). They have also engaged in capacity-building initiatives for institutions in the natural resource sector and made natural resource management a top priority on the local and international agenda. Notwithstanding this laudable progress, the practical impact of the drivers on managing natural resources revenue and thereby reducing the resource curse and Dutch disease has largely remained mixed (Humphreys et al. 2007; Lederman and Maloney 2007; Collier 2011; Gaille 2011).

It is instructive to note that in spite of the progress made, there are still challenges (such as corruption, poor record keeping, and general lack of transparency) to the realization of the three initiatives: the Kimberley Process Certification Scheme (KPCS), Publish What You Pay (PWYP), and the Extractive Industries Transparency Initiative (EITI) (World Bank 2008a; ECA 2009; Obeng-Odoom 2012).

Natural resources governance in Africa: Some vulnerabilities

The natural resources sector's value chain consists of five stages (Humphreys et al. 2007; Dunning 2008): (1) award of contracts, (2) monitoring of operations, (3) collection of taxes and royalties, (4) distribution of revenue, and (5) utilization in sustainable projects.

The chain is governed by laws and regulation to ensure maximum efficiency and effectiveness for favorable beneficial outcomes. Even though natural resource-rich African countries have designed laws and regulations to manage their natural resources, these laws and regulations have been considered "more investment friendly…and in line with international best practices in the industry and as well take cognizance of relevant stakeholder views" (World Bank 2008b: 32). They liberalize and deregulate the sector. However, they also reinforce the point that "In hindsight, and in view of current high mineral prices, some of the natural resource codes then adopted and some of the agreements negotiated may have been overgenerous to foreign investors" (UNCTAD 2007: 161). For instance, in spite of the several reforms of the legal framework of the Ghanaian and Nigerian mining and oil sectors, respectively, the legal framework governing both sectors has been found to be extremely investor friendly and deficient in some areas (Ayee et al. 2011; Gboyega et al. 2011).

Studies on natural resources in Africa highlight vulnerabilities in sector governance along the value chain and explain why it has been difficult to implement best or second best welfare-enhancing policies. The sector is vulnerable to rent-seeking activities because of certain characteristics such as the requirement for large initial capital expenditures, lack of choice in location, the sudden wealth and easy money image, the local nationals' previous experience with MNCs, the particular sense of entitlement that local people have with respect to the wealth generated, and the high level of government regulation (Barma et al. 2012).

There are sector management vulnerabilities in the regulation and award of leases, in revenue collection and administration, and, eventually, in the way budget procedures secure sustainability in revenue reinvestment. Contracts are often subject to strong confidentiality clauses by the MNCs, governments, investors, and banks involved. Corruption is only a part of the explanation. Governments argue that they cannot make all details of the extractive industries public and that they have limited influence on companies. Countries also compete for the scarce managerial and technical skills needed for resource extraction. Yet, shortages of legal and negotiation skills play a major role in driving down tax revenue from natural resources (Collier 2011).

Despite the strong focus on development by many governments, there are still incentive problems in several of the institutions involved in natural resources governance in Africa. An excessive centralized policy-making process, a powerful executive president, strong party loyalty, political patronage, lack of transparency, and weak institutional capacity at political and regulatory levels have greatly contributed to the extractive industries' inadequate flow of net benefits. Consequently, it has been argued that the net benefit of natural resources is likely to be improved with appropriate reforms in governance (Ayee et al. 2011; Gboyega et al. 2011; Barma et al. 2012).

Admittedly, among potentially important values from the industry, other than direct revenue, are transfers of technological and organizational skills and CSR. Multinationals' home country regulations or standards, accounting rules, production technology, and procurement procedures may also help strengthen the sector. Consequently, arguments to downplay the importance of the natural resources to the national interest need to be considered in the light of these perspectives (Eden and Lenway 2001; Obeng-Odoom 2012).

Nevertheless, the economic importance of the natural resources is not adequately matched by forward and backward linkages to other economic growth-promoting activities (Gaille 2011; UNDP 2011). A net benefit assessment requires estimations of benefits from natural resources, including royalties and taxes, infrastructure, technology transfers, and employment, as well as their multiplier effects and how they compare to costs, such as environmental consequences, health problems, cultural difficulties, and loss of agricultural land. The paradox of natural resources wealth is that many of the countries endowed with such resources also have a high level of poverty and experience what is commonly referred to as the "resource curse" (Barma et al. 2012; Obeng-Odoom 2012).

Two reasons are adduced for the paucity of tax revenue from the natural resources sector. First, the contracts with MNCs are often unfavorable for African governments, with calls for renegotiation so far difficult in the light of the stringent contracts. Second, the level of corruption in the sector is generally enormous, with international dimensions (World Bank 2006; Lederman and Maloney 2007; UNDP 2011: Barma et al. 2012).

Capacity development challenges

A number of institutional capacity challenges have been identified in the natural resources sector in Africa (Ross 1999; 2001; Rosser 2006; IMF 2010).

First, the capacity of the legislature in Africa to act as a countervailing force over the executive and understand the complexity of natural resource legislation has been questioned (UNECA 2009; Ayee et al. 2011; Gboyega et al. 2011; UNDP; 2011). Apart from passing the annual budget, the legislature is responsible for ratification of natural resources leases, contracts, and stabilization agreements. However, the performance of these responsibilities is subject to executive influence; thus, the checks and balances intended to secure independent control by the legislature in African countries are rendered dysfunctional. Natural resources leases and other agreements with the companies are first brought to government agencies before being ratified by the legislature and then awarded to the companies. However, supervision by the legislatures and their committees in Africa is not effective because of their weak capacity and the overbearing influence of the executive (UNECA 2009; Ayee et al. 2011; Gboyega et al. 2011; Barman et al. 2012).

Second, weak bureaucratic capacity prevents regulatory institutions from performing their functions effectively. This problem can be identified in almost all ministries, departments, and agencies in Africa (AfDB 2005). This said, the lack of capacity is a function of the lack of political incentives to meaningfully invest in sector reforms rather than a quantitative indicator of a government's scarce human and material resources. Tackling poor regulatory capacity in this arena has not been a consistent priority. As a result, the natural resources industry is not being regulated effectively, essential analyses are not being undertaken, and most policy proposals are accepted without sufficient understanding of their implications. Following from this, capacity is lacking in the implementation of policies, programs, and projects. This general lack of capacity has not been addressed sufficiently at the political level (UNDP'2011).

Third, transfer pricing has negatively affected revenue collections in some African countries. Most MNCs in the natural resources sector operate internationally and have extended dealings with affiliated companies, which increases opportunities for transfer pricing and potentially lowers the tax liability. This further complicates the task of tax administration and creates a challenge that requires specific skills. Most African tax laws have legal provisions to address the issue, but these provisions are insufficient (Brautigam et al. 2008; Barman et al. 2012).

Fourth, the generous concessions granted by governments to MNCs in the natural resources sector cannot be altered even when the conditions in which they were signed change substantially or unexpectedly. Therefore, attempts at renegotiation have not only been contentious but have reflected institutions' weak capacity to meaningfully engage the MNCs, particularly in the face of the resources at their disposal. Royalties and tax concessions often are frozen by an investor-friendly stabilization clause for a set period of time. Higher prices will not necessarily imply a proportional increase in state revenue to mineral-rich African countries (World Trade Organization 2010; Gaille 2011).

Fifth, governments' capacity to deal with the domestic side of natural resource management—for example, effective distribution or use of the revenue generated from natural resources to prevent conflict—has been questioned. Proper natural resource management also involves accountability with regard to how these resources and wealth are used. However, in Africa, poor resource management is a problem in many resource-rich countries in Africa (Onigbinde 2008). For instance, Nigeria is the world's 11th largest oil exporter and derives immense wealth from annual oil production and trade, but it continues to suffer from poor resource management. This is manifest specifically in the Niger Delta region, Nigeria's largest oil-producing region but the poorest because it has not benefited from the oil wealth derived from oil production (Ikelegbe 2006; Obi 2010).

These capacity challenges point to one fact: unlike the government, MNCs have the benefit of the best accounting and legal resources. Moreover, they have been in the countries for a long time through different administrations, and have come to know the system well, and their sheer understanding of how things work creates a competitive advantage. This imbalance may have influenced negotiations between the governments and the MNCs (Ayee et al. 2011 Gaille 2011; Gboyega et al. 2011).

Capacity development opportunities

In the midst of capacity challenges, however, there are capacity opportunities given the renewed interest in curbing the natural resource curse and making natural resources welfare enhancing for host governments. They include the following:

a) The World Bank capacity initiatives

In 2009, the World Bank launched its Natural Resources and Environmental Governance (NREG) Program to assist African countries in managing their natural resources. The mission of the program, which is ongoing, is to improve transparency in systems and procedures for natural resource management, which could lead to more effective enforcement and improved collection of revenue in the sector. In May 1998, the G8 launched an action program on forests, which gives high priority to eliminating illegal logging and illegal timber trade. The action program seeks to complement actions undertaken at regional and international levels and states the G8's commitment to identifying actions in both producer and consumer countries (World Bank 2009). In Ghana, for instance, some of the World Bank's initiatives have led to log tracking system in the forestry sector, preparation of "Social Responsibility Guidelines for Mining Companies in Mining Communities" for the mining sector, and development of a draft Strategic Environmental Assessment on oil and gas (World Bank 2012).

b) International Monetary Fund (IMF) capacity initiatives

In May 2011, the IMF launched its Topical Trust Fund on Managing Natural Resource Wealth, envisioned as an effective vehicle for coordinating donors' capacity-building initiatives in managing natural resource wealth. It has provided US$25 million over five years to scale up technical assistance to low-income and lower-middle-income countries endowed with oil, gas, and minerals to help them deal with associated economic policy challenges. Specifically, it aims to build macroeconomic policy capacities so that countries can avoid the resource curse, and assist countries in getting a fair share of their natural resource wealth so they can invest and spend it wisely. It has concentrated on capacity building in five areas (IMF 2010).

c) Revenue Watch Institute (RWI) capacity-building initiatives

These efforts help societies examine every stage of the development of oil, gas, and minerals, from the decision of beginning exploration through organizing production, establishing revenue management, and designing and implementing policies for spending and economic development. A mainstay of RWI's work is developing the capacity of civil society. It provides financial and technical training and support to more than 50 partner organizations on every aspect of oil, gas, and mining; has provided technical assistance to governments in drafting mining and oil laws and in improving revenue management; and has spearheaded the global campaign to develop global standards for transparency and accountability in the minerals sector (Revenue Watch Institute 2006).

d) United Nations University-Institute for Natural Resources in Africa (UNU-INRA)

Established in 1986, UNU-INRA's work centers on Africa's two most important endowments: its human and natural resources. It aims to strengthen capacities at universities and other national institutions to conduct research and produce well-trained individuals with the ability to develop, adapt, and disseminate technologies that promote the sustainable use of the continent's natural resources. It has also established operating units in the following five countries, through which some of its major activities are undertaken:

- University of Cocody, Abidjan, Côte d'Ivoire (social, economic, and policy analysis related to natural resource management);
- University of Yaoundé I, Yaoundé, Cameroon (use of geoinformatics and applications of computer technology to natural resource management);
- University of Zambia, Lusaka, Zambia (soil fertility and mineral resources);
- University of Namibia, Windhoek, Namibia (marine and coastal resources); and
- Institute for Food Technology (ITA) of the Ministry of Mines and Industry in Dakar, Senegal (processing of agricultural and local food products, food quality, and food technology) (UNU-INRA 1994).

e) Capacity building by universities

Given the importance of natural resources, most universities in Africa have either established departments of natural resource management or introduced programs mainly at the graduate level on various aspects of natural resource management. For instance, the University of Malawi's Bunda College established the Department of Natural Resources Management in 1999 to advance and promote theoretical and practical knowledge in natural resources and environmental management through teaching, research, outreach, and consultancies with the aim of conserving natural resources and the environment.

f) Advocacy of civil society organizations (CSOs)

There are several advocacy CSOs involved in the natural resource sector in all African countries. They have created space for discussion of natural resources issues that have fed directly or indirectly into public policy, capacity-building interventions, and/or initiatives of local and international organizations (Collier 2007; Ayee et al. 2011).

g) Learning from the experiences of countries within and outside Africa

The experience of Botswana will be instructive here given that it has put in place a whole range of policies that have worked together and been supported by effective institutions as well as reinvested the proceeds of natural resources through the use by the government of a Sustainable Budget Index (Leith 2005: 120). The experiences of the United States, Canada, Norway, Oman, Qatar, Indonesia, and Malaysia in natural resource management point to the shared goals of preserving social stability, accelerating economic growth, and creating credible and stable groups of technocrats willing to engage and influence political leaders. Furthermore, strong constituencies outside the natural resource sector were considered and listened to in the management of the proceeds of natural resources. Examples are fisheries in Norway, agriculture in Indonesia, and traditional chiefs and cattle owners in Botswana (Collier 2007; Torvik 2009; UNDP 2011).

It is too early to assess the impact of some of these capacity-building interventions because of their long period of gestation, maturity, sustainability, commitment, and the capacity to learn from experience.

The way forward: Policy recommendations for good natural resource governance and management in Africa

The overall impact of natural resources on Africa's development is mixed, and public discontent has intensified as the sector is perceived as generating low net economic and social returns. Succeeding governments have shown a commitment to addressing some of these issues, but action has been slow, piecemeal, and lacking a holistic approach. The political commitment to decisively reform the sector has been intermittent at best. These issues all point to a lack of good natural resource governance in Africa.

An overarching framework for interrogating natural resource management in Africa must take into account two of the main vulnerabilities in the political economy of natural resource management. The first relates to national sovereignty. The presence of highly professional foreign MNCs and their ability to negotiate with weak government institutions and get better and more generous contracts have echoed concerns over national sovereignty. The second relates to the promotion of a social contract, which imposes an obligation on the part of the government to effectively manage natural resources like mining for the benefit of present and future generations. The model will also deepen support for establishing good governance institutions for natural resource revenue management and to assist countries in negotiating extraction agreements that are both fair and consistent with the development agendas of countries.

This model has the following principles:

- *Transformational and development-oriented leadership.* In a perfect world, a nation's leadership would invest resource proceeds wisely, in a way that maximizes long-term economic development. By either design or an accident of history, resource curse countries in general, and most natural resource-rich African countries in particular,

  > have been beset with a curse of leadership. Leaders representing different regime types and periods of time have shown similar traits in misgoverning their people and misusing their resources. They have also not introduced the right kind of management structures to ensure the use of these resources in a way that benefits their citizens.
  >
  > (Duruigbo 2006: 46)

  The success stories of Norway, Canada, Oman, Qatar, Botswana, Chile, Indonesia, and Malaysia have shown that transformational and development-oriented leadership is a sine qua non for effective management of natural resources because the leaders were able to engage with strong constituencies outside the natural resource sector in the management of natural resource proceeds (UNDP 2011).
- *Development of independent, accountable, and transparent institutions* that can help the government manage the proceeds from natural resources and negotiate meaningfully with MNCs. As Acemoglu et al. (2002) showed in the case of Botswana, quality and competent institutions with clear mandates are needed to make decisions that benefit the welfare of the countries and to enforce transparency and accountability. Strengthening public institutions could generate a political environment less prone to conflict and more efficient in managing public spending. The natural resources sector's enclave character has created public interest in the governments' establishing of legal and institutional structures for regulation that promote accountability and industrial development (Ayee et al. 2011; Barman et al. 2012).
- *Successful natural resources regulation* requires the recognition that the problems of the resource curse are political in nature and need to be tackled at the political level. State elites might have incentive to weaken the very institutions that they have created, and therefore will emphasize the importance of creating institutions supported and

overseen by a dense network of diverse stakeholders (Dunning 2008). A political approach recognizes that policy failures, which result in the resource curse, the "Dutch disease," and other such pathologies associated with natural resource wealth, are not always the result of naivety or lack of capacity on the part of policy-makers. State elites might benefit in the short term from such situations. A framework for political analysis does not assume that all will be solved well if policy-makers and politicians simply know what to do and have sufficient capacity to implement technically sound policies. Instead, it delves into underlying interests and incentives of state actors and puts them at the forefront when constructing policies and strategies (World Bank 2011; Ayee et al. 2011).

- *Deepening capacity-building interventions* involves augmenting skills and knowledge through training, providing technical advice, and enhancing genuine community engagement in all aspects from planning to on-the-ground action. Therefore, capacity building should foster the transfer of technology and technical capacity, social cohesion with communities, and development of human and social capital. Capacity building should be based on the principles of trust, reciprocity, and norms of action and should manage the often-adversarial relationship between the MNCs and local communities (Gelb and Grasmann 2010; UNDP 2011).

- *Learning from success stories.* African countries can learn a lot from the success stories of the United States, Canada, Norway, Botswana, Chile, Indonesia, and Malaysia in the management of natural resources. The countries designed and implemented value-for-money policies and programs that enabled them to avoid the resource curse and the Dutch disease (Larsen 2005, 2006). Learning from others or benchmarking oneself against best practices is part of capacity building. It is only non-learning countries that repeat their mistakes ad infinitum, and the onus is on them to learn from those countries that have performed creditably in the management of natural resources and whose success stories demonstrate how natural resources can become welfare enhancing. This is because "the most interesting aspect of the paradox of plenty is not the average effect of natural resources, but its variation. For every Nigeria or Venezuela there is a Norway or a Botswana" (Torvik 2009: 241).

- *Governments should advocate resource nationalism* by limiting the operations of MNCs and asserting greater national control over natural resource exploitation and development.

References

Acemoglu, D., Johnson, S., and Robinson, J. (2002). "An African Success Story: Botswana," *CEPR Discussion Paper* 3219, London: Centre for Economic Policy Research.

Adamolekun, L. (1999). "Governance Context and Reorientation of Government," in L. Adamolekun (ed.), *Public Administration in Africa: Main Issues and Selected Country Studies*, Boulder, CO: Westview. Chapter 1: 1–16.

AfDB (African Development Bank). (2005). *African Development Report 2005: Public Sector Management in Africa*, Oxford: Oxford University Press.

AfDB. (2010). *African Economic Outlook 2010*, Tunis: OECD/African Development Bank.

Akabzaa, T.M., Seyire, J.S. and Afriyie, K. (2007). *The Glittering Façade: Effects of Mining Activities on Obuasi and Its Surrounding Communities*, Accra: Third World Network-Africa.

Ake, C. (1981). *A Political Economy of Africa*, Essex: Longman.

Alden, C. and Davies, M. (2006). "A Profile of the Operations of Chinese Multinationals in Africa," *South Africa Journal of International Affairs*, 13(1) (Summer/Autumn): 83–96.

Amin, S. (1972). "Underdevelopment and Dependence in Black Africa: Origins and Contemporary Forms," *Journal of Modern African Studies*, 4: 503–521.

Apter, D. and Goodman, L. F. (eds.) (1976). *The Multinational Corporation and Social Change*, New York: Praeger.

Ariweriokuma, S. (2009). *The Political Economy of Oil and Gas in Africa*, London/New York: Routledge.

Auty, R. (1993). *Sustaining Development in Mineral Economies: The Resource Curse Thesis*, London: Routledge.

Auty, R. (2001a). "The Political State and the Management of Mineral Rents in Capital-Surplus Economies: Botswana and Saudi Arabia," *Resources Policy*, 27: 77–86.

Auty, R. (ed.) (2001b). *Resource Abundance and Economic Development*, London: Oxford University Press.

Ayee, J.R.A., Soreide, T., Shukla, G.P., and Minh Le, T. (2011). "Political Economy of the Mining Sector in Ghana," *World Bank Policy Research Working Paper* (July) WPS5730: 1–48.

Babu, S.C. (2000). "Capacity Strengthening in Environmental Natural Resource Policy Analysis: Meeting the Changing Needs," *Journal of Environmental Management*, 58: 1–17.

Barma, N.H., Kaiser, K., Minh Le, T., and Vinuela, L. (2012). *Rents to Riches? The Political Economy of Natural Resource-Led Development*, Washington, DC: World Bank.

Brautigam, D. (1996). "State Capacity and Effective Governance," in B. Ndulu and N. Van der Walle (eds.), *Agenda for Africa's Economic Renewal*, Washington, DC: Translations Publishers for the Overseas Development Council. Chapter 2.

Brautigam, D., Fjeldstad, O.-H., and Moore, M. (eds.) (2008). *Taxation and State-Building in Developing Countries: Capacity and Consent*, Cambridge: Cambridge University Press.

Campell, B. (ed.) (2009). *Mining in Africa*, New York: Pluto Press.

Collier, P. (2007). *The Bottom Billion: Why the Poorest Countries Are Failing and What Can Be Done about It*, New York: Oxford University Press.

Collier, P. (2011). "Managing the Exploitation of Natural Assets in Sub-Saharan Africa," AERC Senior Policy Workshop, Maputo, March 2011.

Collier, P and Hoeffler, A. (2002). "On the Incidence of Civil War in Africa," *Journal of Conflict Resolution*, 46(1): 13–28.

Custers, R. and Matthysen, K. (2009). *Africa's Natural Resources in a Global Context*, Antwerp: IPIS.

DeLancey, V. (2001). "The Economies of Africa," in A.A. Gordon and D.L. Gordon (eds.), *Understanding Contemporary Africa*, 3rd edition, Boulder, CO, and London: Lynne Rienner. Chapter 5.

Doppelhofer, G., Miller, R., and Sala-i-Martin, X. (2000). "Determinants of Long-Term Growth: A Bayesian Averaging of Classical Estimates Approach," Working Paper, NBER.

Drucker, P. (1974). "Multinational Corporations and Developing Countries: Myths and Realities," *Foreign Affairs*, 53 (October): 121–134.

Dunning, J. (1993a). "Governments and Multinational Enterprises: From Confrontation to Cooperation?" in L. Eden and E. Potter (eds.), *Multinationals in the Global Political Economy*, London: Macmillan.

Dunning, J. (1993b). *Multinational Enterprises and the Global Economy*, Reading, MA: Addison-Wesley.

Dunning, J. and Lundan, S.M. (2008). *Multinational Enterprises and the Global Economy*, Northampton, MA: Edward Elgar.

Dunning, T. (2008). *Crude Democracy: Natural Resource Wealth and Political Regimes*, Cambridge: Cambridge University Press.

Duruigbo, E. (2006). "Permanent Sovereignty and Peoples' Ownership of Natural Resources in International Law," *International Law Review*, 33: 44–62.

Dwivedi, O.P. and Nef, J. (1982). "Crises and Continuities in Development Theory and Administration: First and Third World Perspectives," *Public Administration and Development*, 2(1): 59–77.

ECA (Economic Commission on Africa). (2009). *African Governance Report II*, Oxford: Oxford University Press.

Eden, L. and Lenway, S. (2001). "Multinationals: The Janus Face of Globalization," *Journal of International Business Studies*, 32(3): 383–400.

Englebert, P. (2000). *State Legitimacy and Development in Africa*, Boulder, CO: Lynne Rienner.

Ergas, Z. (ed.) (1987). *The African State in Transition*, New York: St. Martin's Press.

Fatton, R. Jr. (1992). *Predatory Rule: State and Civil Society in Africa*, Boulder, CO: Lynne Rienner.

Forstater, M., Zadek, S., Guang, Y., Yu, K., Hong, C.X., and Geo, M. (2010). "Corporate Responsibility in African Development: Insights from an Emerging Dialogue," Chinese Academy of Social Sciences, *Working Paper of the Corporate Social Responsibility Initiative*, No. 60 (October).

Frimpong-Ansah, J.H. (1992). *The Vampire State in Africa: The Political Economy of Decline*, Trenton, NJ: Africa World Press.

Gaille, S. (2011). "Mitigating the Resource Curse: A Proposal for Microfinance and Educational Lending Royalty Law," *Energy Law Journal*, 32: 81–96.

Gboyega, A., Soreide, T., Minh Le, T., and Shukla, G.P. (2011). "The Political Economy of the Petroleum Sector in Nigeria," *World Bank Policy Research Working Paper*, (August) WPS5779: 1–48.

Gelb, A. and Grasmann, S. (2010). "How Should Oil Exporters Spend Their Rents," *CGD Working Paper*, No. 221. Washington, DC: Center for Global Development.

Gordon, D.L. (2001). "African Politics," in A.A. Gordon and D.L. Gordon (eds.), *Understanding Contemporary Africa*, 3rd edition. Boulder, CO, and London: Lynne Rienner. Chapter 4.

Gylfason, T. (2001). "Natural Resources, Education and Economic Development," *European Economic Review*, 45: 847–859.

Gylfason, T. (2004). "Natural Resources and Economic Growth: From Dependence to Diversification," Working Paper, CEPR.

Gylfason T. and Zoega, G. (2006). "Natural Resources and Economic Growth: The Role of Investment," *World Economy*, 29(8): 1091–1115.

Herbst, J. (2000). *States and Power in Africa: Comparative Lessons in Authority and Control*, Princeton, NJ: Princeton University Press.

Hodder-Williams, R. (1984). *An Introduction to the Politics of Tropical Africa*, London: George Allen and Unwin.

Humphreys, M., Sachs, J.D., and Stiglitz, J.E. (eds.) (2007). *Escaping the Resource Curse*, New York: Columbia University Press.

Hyden, G. (1983). *No Shortcuts to Progress: African Development in Perspective*, London: Heinemann.

Ikelegbe, A. (2006). "The Economy of Conflict in the Oil Rich Niger Delta Region of Nigeria," *African and Asian Studies*, 5(1): 23–55.

IMF (International Monetary Fund). (2010). "IMF Launches Trust Fund to Help Countries Manage Their Natural Resource Wealth," Press Release No. 10/497, December 16, 2010, Available at: http://www.imf.org/external/np/sec/pr/2010/pr10497.htm. Accessed September 30, 2012.

Jackson, R.H. and Rosberg, C.J. (1982). *Personal Rule in Black Africa*, Berkeley: University of California Press.

Jeffries, R. (1993). "The State, Structural Adjustment and Good Government in Africa," *Journal of Commonwealth and Comparative Studies*, 31(1) (March): 20–35.

Killick, T. (1978). *Development Economics in Action: A Study of Economic Policies in Ghana*, London: Heinemann.

Larsen, E.R. (2005). "Are Rich Countries Immune to the Resource Curse? Evidence from Norway's Management of Its Oil Riches," *Resource Policy*, 30: 75–86.

Larsen, E.R. (2006). "Escaping the Resource Curse and the Dutch Disease? When and Why Norway Caught Up With and Forged Ahead of Its Neighbours," *American Journal of Economics and Sociology*, 65(3): 605–640.

Lederman, D. and Maloney, W.F. (eds.) (2007). *Natural Resources: Neither Curse nor Destiny*, Washington, DC: World Bank and Stanford University Press.

Leith, J.C. (2005). *Why Botswana Prospered*, Montreal: McGill-Queen's University Press.

Leonard, H.J. (1980). "Multinational Corporations and Politics in Developing Countries," *World Politics*, 32(3) (April): 454–483.

Lesourne, J. and William, R.C. (2009). *Governance of Oil in Africa: Unfinished Business*, Paris: IFRI.

Levy, B. (2004). "Governance and Economic Development in Africa: Meeting the Challenge of Capacity Building," in B. Levy and S. Kpundeh (eds.), *Building State Capacity in Africa: New Approaches, Emerging Lessons*, Washington, DC: World Bank. Chapter 1.

Leys, C. (1976). "The 'Overdeveloped' Post-Colonial State: A Reevaluation," *Review of African Political Economy*, 5(January–April): 39–48.

Macleod, F. (2012). "Platinum Belt's Social Ills Fuel Unrest," *Mail and Guardian*, August 17–23, 2012.

Mafeje, A. (1977). "Neo-Colonialism, State Capitalism or Revolution," in P. Gutkind and P. Waterman (eds.), *African Social Studies*, London: Heinemann. Chapter 3.

Mehlum, H., Moene, K.O. and Torvik, R. (2006). "Institutions and the Resource Curse," *Economic Journal*, 116(5): 1–20.

Mildner, S.-A., Lauster, G., and Wodni, W. (2011). "Scarcity and Abundance Revisited: A Literature Review on Natural Resources and Conflict," *International Journal of Conflict and Violence*, 5(1): 155–172.

Moran, T.H. (1978). "Multinational Corporations and Dependency: A Dialogues for Dependentistas and Non-Dependentistas," *International Organization*, 31(1) (Winter): 79–100.

Myrdal, G. (1968). *The Asian Drama: An Enquiry into the Poverty of Nations*, New York: Random House.

Obeng-Odoom, F. (2012). "Problematizing the Resource Curse Thesis," *Development and Society*, 42(1): 1–29.

Obi, C. (2010). "Oil Extraction, Dispossession, Resistance, and Conflict in Nigeria's Oil-Rich Niger Delta," *Canadian Journal of Development Studies*, 30(1/2): 219–236.

Onigbinde, D. (2008). "Natural Resource Management and Its Implications on National and Sub-Regional Security: The Case of the Niger Delta," *Occasional Paper*, No. 22: 1–23, Kofi Annan International Peacekeeping Training Centre (KAIPTC).

Ozoigbo, B.L. and Chukuezi, C.O. (2011). "The Impact of Multinational Corporations on the Nigerian Economy," *European Journal of Social Sciences*, 19(3): 380–387.

Pritchard, W. (2010). "Taxation and State Building: Towards a Governance Focused Tax Reform Agenda," *IDS Working Paper*, 341: 1–55.

Ramamurti, R. (2001). "The Obsolescing Bargaining Model? MNC-Host Developing Country Relationships Revisited," *Journal of International Business Studies*, 31(1): 23–39.

Ramaphosa, C. (2012). Interview on SABC Channel 6 Morning Talk, September 20, 2012.

Revenue Watch Institute. (2006). "The Evolution of Revenue Watch Institute," Available at http://revenuewatchinstitute.org. Accessed September 3, 2012.

Riggs, F. (1964). *Administration in Developing Countries: The Theory of Prismatic Society*, Boston: Houghton Mifflin.

Robinson, J. (ed.) (1979). *The International Division of Labour and Multinational Companies*, London: Saxon House, Teakfield Ltd.

Rodney, W. (1972). *How Europe Underdeveloped Africa*, London: George Allen and Unwin.

Ross, M. (1999). "The Political Economy of the Resource Curse," *World Politics*, 51: 297–322.

Ross, M. (2001). "Does Oil Hinder Democracy?" *World Politics*, 53(April): 325–361.

Rosser, A. (2006). "Escaping the Resource Curse," *New Political Economy*, 11(4): 557–570.

Sachs, J.D. and Warner, A.M. (2001). "The Curse of Natural Resources," *European Economic Review*, 45: 827–838.

Sala-i-Martin, X. (1997). "I Just Ran Two Millions Regressions," *American Economic Review*, Papers and Proceedings, 87(2): 178–183.

Salami, M.B. (2001). "Environmental Impact Assessment Policies, their Effectiveness or Otherwise for Mining Sector Environmental Management in Africa," in Third World Network (ed.), *Africa, Mining Development and Social Conflict in Africa*, Accra: Third World Network.

Sklar, R. (1975). *Corporate Power in an African State: The Political Impact of Multinational Mining Companies in Zambia*, Berkeley: University of California.

Stevens, P. (2008). "National Oil Companies and International Oil Companies in the Middle East: Under the Shadow of Government and the Resource Nationalism Cycle," *JWELB*, 1(1): 12–20.

Strauss, M. (2000). "The Growth and Natural Resource Endowment Paradox: Empirics, Causes and the Case of Kazakhstan," *Fletcher Journal of Development Studies*, 16: 1–28.

Synder, R. (2006). "Does Lootable Wealth Breed Disorder? A Political Economy of Extraction Framework," *Comparative Political Studies*, 39(8): 943–968.

Tandon, Y. (1979). *In Defence of Democracy*, Dar es Salaam: Dar es Salaam University Press.

Tangri, R. (1999). *The Politics of Patronage in Africa: Parastatals, Privatization and Private Enterprise*, London: James Currey.

Torvik, R. (2009). "Why Do Some Resource-Abundant Countries Succeed While Others Do Not?" *Oxford Review of Economic Policy*, 25(2): 241–258.

Turner, M. and Hulme, D. (1997). *Governance, Administration and Development: Making the State Work*, London: Palgrave Macmillan.

UNCTAD. (2007). *World Investment Report 2007*, Geneva: UNCTAD.

UNDP. (2011). *Managing Natural Resources for Human Development in Low Income Countries*, Working Paper 2011–002 December.

UNECA. (2009). *African Governance Report II*, Oxford: Oxford University Press.

UNU-INRA. (1994). "History of the UNU-INRA", Available at http://www.unu-inra.org. Accessed September 11, 2012.

Vernon, R. (1977). *Storm over the Multinationals: The Real Issues*, Cambridge, MA: Harvard University Press.

Ward, H. (2009). "Resource Nationalism and Sustainable Development: A Primer and Key Issues," *International Institute for Environmental Development Working Paper*, 2 (March): 1–23.

Wiig, A. and Kolstad, I. (2010). "Multinational Corporations and Host Country Institutions: A Case Study of Corporate Social Responsibility in Angola," *International Business Review*, 19(2): 178–190.

World Bank. (1997). *World Development Report 1997: The State in a Changing World*, New York: Oxford University Press.

World Bank. (2006). *Where Is the Wealth of Nations—Measuring Capital for the 21st Century*, Washington, DC: World Bank.

World Bank. (2008a). *Implementing the Extractive Industries Transparency Initiative*, Washington, DC: World Bank.

World Bank (2008b). *International Development Association Programme Document for a Proposed Credit in the Amount of SDR8.2 Million (SS$13 Million Equivalent) to the Republic of Ghana for a Natural Resources and Environmental Governance First Development Policy Operation*, Report No. 42787-GH, May 7, 2008.

World Bank (2009). *Second Natural Resources and Environmental Governance Development Policy Operation*, Washington, DC: World Bank.

World Bank (2011). *Doing Business 2011: Making a Difference for Entrepreneurs*, Washington, DC: World Bank.

World Bank (2012). "NREG Program Results in Ghana", Available at http://www.worldbank.org. Accessed September 26, 2012.

World Trade Organization (2010). *World Trade Report 2010: Trade in Natural Resources*. Geneva: World Trade Organization.

Wunsch, J.S. and Olowu, D. (1990). *The Failure of the Centralized State: Institutions and Self-Governance in Africa*, Boulder, CO: Westview Press.

Young, C. (1994). *The African Colonial State in Contemporary Perspective*, New Haven, CT: Yale University Press.

Young, C. (2000). "The Heritage of Colonialism", in J.W. Harbeson and D. Rothchild (eds.), *Africa in World Politics*, Boulder, CO: Westview.

3
Governance of Natural Resource Management in Africa: Contemporary Perspectives

Peter Arthur

Introduction

With escalating energy and raw material demands from the BRIC (Brazil, Russia, India, and China) countries and increasing commodity prices in the world market, there is now renewed global interest in minerals and natural resources throughout Africa (Maconachie 2009: 73). While the control of natural resources such as oil, gas, minerals, forests, and water has the potential to confer great economic and strategic benefits to the countries where they are found (Arthur 2012), there is also increasing evidence that extractive natural resources have not helped developing countries, especially those in Africa, to achieve prosperity and their desired socioeconomic ends. Indeed, for many African countries, such as Nigeria, Sierra Leone, Angola, the Democratic Republic of Congo, and Liberia, the discovery and exploitation of extractive natural resources has only contributed to increasing political instability, conflict, wars, and socioeconomic degradation. Such a state of affairs has spawned what is referred to in the literature as the "resource curse." The notion of the resource curse suggests that countries with large caches of natural resources often perform worse in terms of economic growth, social development, and good governance than other countries with fewer resources. The theory posits that countries depending on oil or other extractive industries for their livelihood are among the most economically troubled, socially unstable, authoritarian, and conflict-ridden in the world (Sovacool 2010: 225). Unsurprisingly, questions about effective and efficient management of natural resource revenues have become increasingly central to local and international efforts. Despite various recommendations, the literature on natural resource management has given little attention to transboundary resource management, the importance of equitable distribution of benefits, and the role of good governance and capacity development initiatives, and this is the gap that this research hopes to fill.

Thus, the chapter starts in the section titled "State of Africa's Natural Resources: Literature Review" with an in-depth literature review, also addressing the existing theoretical arguments of the resource curse. It then turns in the section titled "Natural Resource Management Initiatives" to analyze the various initiatives that have been adopted to promote efficient natural resource management. Among those analyzed are the following transboundary governance arrangements: Southern African Development Community (SADC) Protocol on Shared Watercourses, and the 1999 Nile Basin Initiative (NBI). The role of private sector initiatives such as the Extractive Industries Transparency Initiative (EITI) and the Kimberley Process Certification Scheme (KPCS) in helping ensure that natural resources are better managed by African governments are also discussed. In the section "Fostering Good Governance in Natural Resource Management: Transparency, Accountability, Rule of Law, and Participation," the chapter focuses on the role of good governance as reflected in accountability and transparency measures, and its impact on and contribution to natural resource management. The ability of non-state actors such as the media and civil society to perform watchdog functions, thus ensuring transparency and accountability in resource management, is also analyzed. The final section offers suggestions and recommendations regarding the best approaches to promoting natural resource management in Africa. Specifically, the recommendations are focused on transboundary resource management, and the issue of equitable distribution of benefits in natural resource-rich African countries. Particularly, the positive experience of Botswana with a view to learning from its successfully managed diamond resource is looked at. In addition, the role of good governance as well as the capacity development initiatives that need to be put in place to make natural resources in African countries a blessing as opposed to a "resource curse" are also examined.

State of Africa's natural resources: Literature review

The general expectation and belief is that revenue from natural resources could help promote socioeconomic development in resource-rich countries. Unfortunately, this is not always the case. Studies (Karl 1997; Sachs and Warner 2001; Humphreys et al. 2007) have pointed to the fact that developing countries with abundant natural resources generally experience less than impressive socioeconomic growth and performance when compared with those with fewer natural resources. The negative relationship between extractive natural resources on the one hand and economic development on the other, throughout much of the developing world, is what Karl (1997) termed the "paradox of plenty." According to Karl (1999: 34), the "resource curse" or "paradox of plenty" is explained by the fact that many of the oil-producing countries or petro states are more dependent than other states on a single non-renewable commodity, and the exploitation of that commodity is more capital intensive, more enclave oriented, more centralized in the state, and more rent producing than any other, all of which bodes ill for successful development. Revenue flowing through incapable or corrupt structures would have negative and perverse consequences on a country because the money was put to bad use, encouraged dependence on natural resources, and tended to result in corruption and a mono-industry economy. In particular, oil exports inflated the value of a country's currency and made its other exports uncompetitive. In sum, natural resource extraction created and solidified asymmetries in wealth and increased the income gaps between the rich and poor; this, in turn, contributed to the institutionalization of corruption and enabled oppressive regimes to maintain their political power (Karl 1997). Another argument holds that resource-cursed governments tended to manage natural resources opportunistically rather than strategically. That is, government officials viewed them as mechanisms to create immediate wealth rather than long-term but nonrenewable assets that should be carefully managed over time. Resource depletion therefore tended not to serve the public interest, and instead consolidated wealth among a very narrow class of bureaucrats and managers. Moreover, the immense value of those resources masked mismanagement and inequity (Sovacool 2010: 229).

While the discovery of natural resources does not necessarily lead to economic growth and development, it can have the undesirable effect of triggering or fueling instability, weakening governing institutions, and making society vulnerable to armed conflict (Le Billon 2008: 347). The violent conflict, militant activities, and insurgency permeating the Niger Delta region of Nigeria can be explained by alienation, despair, marginalization, the desire for self-determination and local autonomy, and the lack of opportunities for young people in the region (Obi and Rustad 2011). Indeed, over 85 percent of the region's working population has no connection with the oil and gas industry, with low-wage/low-productivity informal enterprises as their primary source of livelihood (Ahonsi 2011: 29). In addition, environmental degradation, poverty and unemployment, infrastructural underdevelopment, and the lack of political participation and democratic accountability have contributed to the ongoing conflict in the Niger Delta region (Ahonsi 2011). Furthermore, some oil companies operating in the Niger Delta either made direct payments to armed groups or awarded contracts to them ostensibly to provide "security" for oil installations, thereby fueling local conflict dynamics. These multinationals also blamed oil spills on acts of sabotage by oil thieves and militias and played down the intimate relationship and complicity between the companies and the state, symbolized in part by company payments to state security forces guarding oil installations and bribes paid by multinationals to top state officials to secure oil contracts (Obi 2010: 487). The literature further notes that even when conflicts gave way to a fragile peace, control over natural resources and their revenue often remained in the hands of the a small elite from the winning side or were shared among the competing elites. Hence, the risks of violent conflict were found significantly higher in mineral resource-rich countries than in countries with other abundant natural resources such as fertile land (McFerson 2010: 337).

A final prominent theory to emerge in recent times to explain why the existence of natural resources can lead to a "curse" is related to the idea of state failure (de Soysa 2011).[1] The nexus between lootable resource wealth, poor governance, underdevelopment, and conflict might be particularly prominent in failing states where large deposits of alluvial minerals are mined artisanally, and it remained virtually impossible to monitor or regulate their extraction (Maconachie 2009). For de Soysa (2011), where there is state failure, having abundant resources might directly lead to large-scale civil and international violence, regardless of other factors. Natural resources may act as a "honeypot" that attracts predation. Resource wealth also allowed groups to overcome financial constraints of organizing large-scale violence by forming on the basis of looting natural wealth. Thus, when there is a "failed state," conflict could be self-financing due to the high value of "capturable" natural resources, such as diamonds (Sierra Leone, Angola) and oil (Nigeria).

Given the overlap between state failure and violent resource wars, some commentators have argued that the panacea for resource curse could be found in the promotion of good governance. Collier (2007) noted that abundant natural resources and the absence of good governance often led to greed and helped turn a "resource into a curse." Poor or bad governance could affect equitable distribution of resources, especially when, as in the case of the Niger Delta region of Nigeria, a particular region's inhabitants believe that their resources are being stolen by political elites with little benefits to the resource-host communities. In Nigeria, the 13th largest producer of petroleum in the world, oil and gas accounts for 80 percent of all government revenue, 95 percent of exports, and 90 percent of foreign exchange earnings. Oil production has generated billions of dollars in government revenue, but the country's average gross domestic product (GDP) in the past decade was less than its GDP in the 1960s, and the poverty rate increased from 27 percent in 1980 to 70 percent in 1999 (Omorogbe 2006: 44 cited in Sovacool 2010: 233).

These governance-related problems are exacerbated by the fact that there are many examples in Africa, as epitomized in the Great Lakes and Nile River basin, where natural resources such as water cut across or traverse the national boundaries, making their management even more challenging and increasing the real possibility of large-scale violent conflict (Swatuk 2012: 85). The dominant narrative of water and conflict in Africa tended to elide these factors: that is, a difficult hydrology made more so by climate change, combined with watercourses shared by two or more states, each of which is interested in harnessing more water for agriculture, industry, and other cities, ineluctably leads to competition for an increasingly scarce resource (Swatuk 2012: 87).

Despite evidence in the literature, some of the claims of the resource curse theorists have been challenged more recently (Obi 2010: 484) because the resource-conflict link is probably more complex than is conceptualized in the scientific mainstream (Basedau and Lay 2009: 758). Contrary to the aspect of rentier state theory that posited that resource-rich states were weak, corrupt, authoritarian, and therefore susceptible to conflict (Obi 2010: 488), the argument was made that "governments use revenue from abundant resources to buy off peace through patronage, large-scale distribution policies and effective repression"[2] (Basedau and Lay 2009: 758).

The notion of resource curse has also been criticized from a methodological and econometric perspective. These criticisms particularly concerned the trade-based proxies (such as the share of primary product export) traditionally used to measure natural resource abundance. From this perspective, the disappointing growth performance of resource-rich countries was related to macroeconomic policies rather than to natural resources. Thus, the resource curse appeared to be due to econometric and measurement fallacies, while resource wealth may represent an important factor for economic development (Daniele 2011: 547–548).

Finally, examining the relationship between oil and conflict, for example, Fearon (2005) argued that oil predicted civil war risks not because it provided an easy source of rebel start-up finance, but probably because oil producers have relatively low state capabilities that stemmed from their level of per capita income as well as weak military and institutional structures capable of effectively repressing the outbreak of armed insurrection, and because oil made state or regional control a tempting prize.

Such criticisms and challenges against the protagonists of the resource curse theory have led them to partly move away from the initial debates over the "greed versus grievance" causal binary. It is therefore unsurprising that much of the emphasis has shifted to issues related to the risk, onset, duration, and intensity of armed conflict in resource-rich countries, and exploration of the links between resource endowment and the viability or capacity of rebel groups (Obi 2010: 484). While the resource curse thesis could be criticized and its core arguments undermined on the grounds of its "prevailing evaluation methodologies and on the basis of measurement errors, incorrect specification of the models and the high probability of spurious correlations" (UNRISD 2007: 12 cited in Obi 2010: 489), it would at the same time be disingenuous to ignore the general ills, economic challenges, and political and social pains that resource-rich African countries face in managing those natural resources. As a result, an issue that often crops up has been the means by which natural resources could be managed to enhance cooperation and contribute to the overall socioeconomic development of countries.

Natural resource management initiatives

a) Transboundary natural resource management

In the efforts to ensure that natural resources like water, forests, and minerals become a blessing as opposed to a curse, numerous steps toward integrated planning and decision-making are taking place across Africa. For example, as Swatuk (2012: 84) has pointed out, the fact that there is no substitute for water as a necessary resource for life, combined with its shared nature across many African states, has led to the suggestion by scholars that water cooperation could have tangible peace and development effects "beyond the river." Within Africa, many countries have recognized the need for cooperative management of transboundary waters, and this has found expression in numerous bilateral and multilateral agreements and transboundary river basin organizations (TRBOs). An example of regional cooperation among member countries and implementing agencies that has contributed to effective and successful resource management is the SADC Protocol on Shared Watercourses (Mirumachin and Van Wyk 2010: 30). Signed by 13 SADC Heads of State in Windhoek, Namibia, on August 7, 2000, the protocol, which is an instrument of international water law to jointly manage water in the Orange River basin,[3] has fostered closer cooperation between the SADC states for the coordinated management, protection, and utilization of shared watercourses through the establishment of river basin organizations (Heyns et al. 2008). Bilateral agreements and, eventually, a basin-wide multilateral agreement to establish the Orange-Senqu River Commission (ORASECOM) not only reduced the potential for conflict, but also contributed to regional integration, socioeconomic development, poverty alleviation, and the protection of vital ecosystems in the region (Heyns et al. 2008: 371–377).

Similar to the SADC Protocol, there is also the 1999 Nile Basin Initiative (NBI), which was launched by nine riparian countries with assistance from the international community to not only offset the negative trends in resource use and the conflict potential of Egyptian hydro-hegemony (Swatuk 2012: 99), but also ensure cooperation and economic integration, sustainable resource development, and security (Teshome 2008). The NBI was followed by the launching in 2006 of the Nile River Basin Commission (NRBC) to foster cooperation and sustainable, equitable, and peaceful use of water resources of the Nile River (Kagwanja 2007: 311). As Swatuk (2012: 99–100) posts, the basic principles of the NBI are transboundary environmental action, regional power trade, efficient water use for agricultural production, and water resource planning and management. Despite its principles, goals, and multitrack strategy with a development focus, which can serve as a model for other transboundary resource management (especially mineral resources that straddle national borders or waters), the NBI has faced a number of challenges. Notable among them are the lack of overall political leadership, mutual suspicions and distrusts among the "upstream" and "downstream" countries on water resource development, political problems among some members of the NBI, the absence of agreement on water allocation among the riparian states that is accepted by all member countries, the lack of a strong legal and institutional framework, poor infrastructure, poverty, inadequate skills, and environmental degradation (Teshome 2008: 37–38). In fact, Swatuk (2012: 100) has pointed out that the Nile basin case illustrated the complexity and difficulty of arriving at a mutually acceptable arrangement for managing transboundary waters, particularly where the downstream state (Egypt) has captured the resource (for which there is no substitute) and is considerably more powerful than all other states in the basin, while the upstream states lack significant human, financial, and other resource capacities.

Notwithstanding the complexities, occasional interstate tensions, and other challenges, there is no denying that the NBI has largely resulted in a multilateral cooperative approach to sharing water and its benefits among members. Kagwanja (2007) argues that the NBI witnessed a shift from antagonism to cooperation among riparian states in the utilization of resources, and that the launch of the NBI and the concomitant NRBC signified the triumph of regionalism over unilateralism in managing and settling conflicting claims over shared water resources.

b) Kimberley Process Certification Scheme (KPCS)

Another approach that has recently emerged regarding how to efficiently manage natural resources has centered on the KPCS, named after its inaugural meeting place in Kimberley, South Africa (Grant and Taylor 2004: 387). Launched in 2000 as a joint government, industry, and civil society initiative and established in 2003, the scheme aimed to prevent the sale of diamonds perceived to have been mined in conflict areas. The idea of the KPCS, which as of August 2012 counted 75 countries as members, including all major diamond-producing, -trading, and -processing countries, was to cut off funding to rebel groups involved in wars against legitimate governments. The KPCS empowers governments to hold companies legally responsible for the statements they make with regard to certificates (Wexler 2010) and, through this process, ensures that "blood diamonds" did not enter the international diamond market. As Wright (2004: 702) noted, the KPCS represented something unique in the annals of international diplomacy. It was the first time that a serious attempt had been made by the international community to address the problem of the illegal exploitation of natural resources. It was also the first time that an international agreement had been negotiated and adopted on the basis of consensus between governments, industry, and civil society acting as equal partners. Finally, by putting violators at risk of market exclusion, the scheme clearly had an influence on exporting states' cost–benefit calculus (Carbonnier et al. 2011: 253).

Despite these laudable goals and achievements, the KPCS has had to contend with a number of challenges. Diamonds, whether rough or polished, are often small, and even larger stones are relatively light.

This makes diamonds easy to smuggle unless customs agents employ x-ray devices (detectable under such scanners) (Grant and Taylor 2004: 396). Thus, dealing with smuggling and other forms of illegitimate trading in natural resources like diamonds remains a challenge. Furthermore, it is unclear what penalties, if any, would be applied to transgressors, aside from possible expulsion from the KPCS. Finally, there were concerns relating to its voluntary and nonbinding nature, weak government oversight, as well as weak controls and monitoring abilities by the actors involved (Grant and Taylor 2004; Wright 2004; Wexler 2010). As Grant and Taylor (2004: 397) point out, the "certificate of origin" issued by governments may be undermined by a lack of transparency within some national diamond industries. They note that the governments of countries such as Angola, Russia, and China were adamant that information on diamond resources—ranging from ownership of private shares in mining joint ventures to precise mine location and production statistics to tax collection procedures and figures—was a matter of "national security" and therefore could not be revealed.

c) Extractive Industries Transparency Initiative (EITI)

Another voluntary code that has gained international recognition is the EITI. Launched in 2002, its underlying assumption was that the opacity that marked business-government relationships in the extractive industry facilitated greed, the mismanagement of natural resource revenue, and the inability to hold governments accountable (Idemudia 2009:10). To address this, the EITI aimed to increase transparency in financial transactions between governments and companies within the extractive industries.

As an initiative that aimed at improving oversight and preventing corrupt practices, such as the diversion of revenue from extractive industries intended for public government accounts into private accounts, the EITI involved the full publication and verification of company payments made to governments and of government revenue received from oil, gas, and mining activities. With payment and revenue information made transparent and publicly available, it became easier to exert pressure on governments for better spending on key basic services such as education and health (Arthur 2012). Moreover, a regular provision of quality information created a system of checks and balances, which assisted in holding companies and governments accountable and resulted in improved economic importance, political stability, and better investment (Maconachie 2009).

It is therefore a welcome development that many African countries have signed up to implement the EITI. While Nigeria in 2007 became the first candidate country with a statutory backing for the implementation of EITI (Idemudia 2009), 22 African countries are now signatories to the EITI, which has committed them to greater resource revenue transparency. The Nigerian government, for example, recognized that improvements in the transparency of petroleum revenue data were needed for the effective management of public resources and to improve the image of Nigeria at home and abroad (Idemudia 2009). Similarly, in Cameroon and Gabon, EITI helped with the monitoring and management capacity of government agencies and provided the platform for civil society organizations (CSOs) to identify challenges in the management of natural resources (EITI Secretariat 2010).

While advocates believed that the EITI held the key to facilitating economic improvement in resource-rich developing countries (Maconachie 2009: 72), it would at the same time be presumptuous to assume that they are the magic bullet for meeting all the challenges associated with resource mismanagement among African countries (Arthur 2012). This is because the EITI has its own unique challenges. First, according to Kolstad and Wiig (2008: 228), the EITI was an initiative that focused on revenue from extractive industries in resource-rich countries. This implied a narrow take on transparency, as only a small section of the public sector was covered. Other parts of the natural resource extraction value chain were not addressed by the initiative, which also failed to address transparency in the use of public resources—the expenditure side, which was clearly the key in many of the corruption-related problems faced by resource-rich countries. Aside from its voluntary nature, another major weakness of the EITI approach was that companies that participated in the initiative were obliged to report payment only in countries of operation that subscribe to EITI (Arthur 2012: 116). Moreover, in Nigeria, for example, the government failed to enact legislation that would have complemented its participation in EITI by making government expenditure at all levels more transparent. The government also failed to pass into law a fiscal responsibility bill to introduce new measures of integrity, transparency, and uniformity of budget making and government expenditure at all levels (Idemudia 2009).

In Angola, little information was available to public about the state budget, with the exception of the enacted budget (de Renzio et al. 2005: 65). Documents were produced primarily for internal purposes, but sometimes they were not produced at all, for lack of adequate capacity, as in the case of in-year reports or audited accounts. Also, in Cameroon, a criticism of the EITI concerned the poor quality of the information provided in the official documents published as part of the EITI. According to Gauthier and Zeufack (2011: 58), only aggregated figures for the country's total oil production were presented, with no details provided for each of the companies' production and payments to the treasury and the National Hydrocarbons Company (Société Nationale des Hydrocarbures, SNH), a public corporation controlled by the presidency. Finally, a country like Botswana, which is often touted as a success story in terms of avoiding the resource curse, was initially reluctant to be part of the EITI because it would have breached its confidentiality agreements with business partners like De Beers. Indeed, the position of the Botswana government regarding the EITI in 2003 was that, although it believed in transparency and accountability, publication of national revenue acquired from extractive industries should be tempered by commercial and political prudence. The Botswana government suggested that in a competitive marketplace, it was not prudent for it to publish all its commercial secrets for the use and best benefit of its competitors. Although Botswana ultimately joined the EITI in 2007, its initial reluctance mirrored the concerns and challenges that have bedeviled the EITI.

Fostering good governance in natural resource management: Transparency, accountability, rule of law, and participation

The argument for good governance as a panacea for resource mismanagement in Africa is that it promotes a society's development of human capabilities and ensures transparency and accountability, each of which contributes to achieving overall development (Maconachie 2009). Alao (2007) has argued that the "governance structure" around extracting and processing resources and managing generated revenue determined whether natural resources would turn out to be a curse or a blessing. It was the case that many developing countries were unable to take full advantage of their natural resources, and the underlying causes of such a state of affairs could be explained by governments' failure to deal with institutional infrastructure and policy-related challenges such as weak existing laws, regulatory support, and technical expertise.

As Wantchekon (2002: 2) argues, when state institutions were weak so that budget procedures either lacked transparency or were discretionary, resource windfalls tended to help consolidate an already-established authoritarian government and generated incumbency advantage in democratic elections. This incited the opposition to resort to political violence in competing for political power, which generated political instability and authoritarian governments. This argument was similar to that of Mehlum, Moene, and Torvik (2006), who demonstrated through regression analysis that the resource curse was likely to occur in countries that lacked strong institutions and good governance. In Nigeria, for example, the absence of strong institutions, political transparency, and accountability, coupled with the high incidence of corruption, have all combined to undermine various levels of governments' ability to put in place policies and programs that would contribute to the overall socioeconomic development of the oil-producing areas (Ite 2004: 5). The state of affairs in Nigeria is not different from the situation in Cameroon. As Gauthier and Zeufack (2011: 27) pointed out, although Cameroon may have captured an estimated 67 percent of its oil rent, only approximately 46 percent of total oil revenue that accrued to the government between 1977 and 2006 may have been transferred to the budget. The remaining 54 percent was not properly accounted for due to poor governance and the lack of a transparent and accountable framework to manage oil revenue.

Similarly, in Angola, within the social sectors, financial resource allocations were biased to elite interests. For example, funding for overseas scholarships and medical education were given to the elites at the expense of the most elementary primary health-care and basic education needs of the Angolan people (McFerson 2009a: 1533). Aside from that, across resource-dependent political economies, rent seeking provided the ruling elites with the means of maintaining their hegemony. Such a state of affairs divided countries between a privileged and wealthy ruling minority, while the rest of society remained impoverished (Arthur 2012: 125). For example, despite being the source of over 90 percent of the oil explored and exported in Nigeria and of foreign exchange earnings, communities in the Niger Delta region of Nigeria, including the nine oil-producing states (Abia, Akwa Ibom, Bayelsa, Cross River, Delta, Edo, Imo, Ondo, and Rivers), generally lack essential social services and infrastructure, and local people are not employed in the oil industries operating in their communities. It is in this regard that Siegle (2005: 48) argued that the whole problem with the resource curse centered on unfairness. This was because a privileged minority benefited extravagantly from their insider status at the expense of the majority. This inequity persisted because those in power were able to take advantage of a lack of public scrutiny to conceal from the public the degree to which they are profiting from these national endowments.

It is therefore unsurprising that McFerson (2009b; 2010: 338) identified the major pillars of good governance as accountability, transparency, the rule of law, and participation, which could help in resource management. While accountability consisted of the capacity to call public officials to task for their actions, especially with regard to how public revenue is mobilized and spent, transparency entailed the low-cost access by citizens to relevant information, particularly on public service access and quality, mobilization of revenue, and allocation of government expenditure.

In addition, transparency of government information is a must for an informed executive branch, legislature, and public at large—normally through the filter of competent legislative staff and capable and independent public media. The rule of law is also critical to provide society and the private economic sector with predictability—which, in addition to formal laws, required regulations and administrative provisions that were clear, known in advance, and uniformly and effectively enforced. Finally, participation by users of services, government employees, other relevant stakeholders, and citizens at large is necessary to design effective government programs, supply the government with reliable information, and provide it with a reality check (McFerson 2010: 338).

The importance of an informed media and civil society in natural resource management

Good governance could be further fostered by having the media and CSOs play important roles in the affairs of a country. As the fourth estate, an objective media remains instrumental in ensuring accountability through its watchdog and monitoring functions. By focusing on and providing information to the general public, the media can enlighten the general public on how revenue from the resource sector, for example, is being disbursed by the government. This process, however, calls for specialized training in the area of natural resource management and extractive reporting, as without the requisite know-how by the media, the information provided to the general public might sound hollow. Thus, having an informed media that is also very knowledgeable about the issues is crucial to ensuring effective resource management and helping the ordinary person or citizens to be educated.

Beyond the role and contribution of the media, effective resource management can be further enhanced if individuals and civil service organizations have the political space for expression (Arthur 2010).

CSOs in various countries have played a vital role in fostering debate to improve transparency and accountability (de Renzio et al. 2005: 59). Thus, African states have the responsibility to accommodate the complaints and demands made by civil societies, as these ultimately serve as important monitoring agents in states with weak institutions (Gonzalez-Vicente 2011: 73). Furthermore, civil society can also play an effective role in engaging transnational investment through direct pressure, such as strikes at the site of investment projects. The prerequisites for the success of this approach are the willingness of state forces to allow protests and the capacity of the companies' local managers to bring about the demanded changes (Gonzalez-Vicente 2011: 72).

This approach has been successfully put into practice in Zambia, where Chinese investors have been very active in recent times. The death of 46 Zambian workers in the Chinese-operated Chambishi mine in 2005 raised the alarms of civil society. This pushed the Chinese mine to increase its social investment, but also ignited a democratic debate about Zambia's partnership with China (Gonzalez-Vicente 2011: 73–74). In addition to using direct pressure as employed in Zambia, civil society can also be involved in efficient resource management by tackling ownership (i.e., shareholders) at the corporation's country of origin. This has been a preferred way in which transnational advocacy networks have successfully internationalized local struggles. Connections between local civil society and transnational advocacy networks are essential to rescale and empower opposition narratives (Gonzalez-Vicente 2011).

The role of civil society groups is also evident in a country like Ghana, where they have been active in advocating the adoption of legislation in line with international best practices. According to Gyampo (2011: 11), as part of a strategy to reclaim and open up the space for the democratic participation of civil society in the consultation process, a preparatory workshop sponsored by Revenue Watch Institute, Oxfam America, Catholic Relief Services, and other organizations was organized to collate civil society views and concerns to feed into the forum. The prime objective of the preparatory meeting was to educate civil society on oil production and oil development issues on the basis of international best practice and to formulate a set of civil society demands on the forum. Also, the meeting served as the rallying point for the few civil society representatives invited to the national forum to carry the voices of many who had been denied the opportunity to participate in the national consultative process (Gyampo 2011: 53).

In particular, the Oil and Gas Platform, a network of approximately 35 CSOs working on oil and gas issues in Ghana, has been very active by engaging in analytical capacity and advocacy (Prempeh and Kroon 2012). Set up in 2008 with the support of Oxfam and the World Bank, the Oil and Gas Platform conducts capacity audits to determine what capacity exists in the oil and gas sector and where it is lacking. Also, it has the intention of focusing on compliance monitoring of petroleum laws and social and environmental monitoring. In addition, given the strategic role civil society is expected to play in promoting accountability and community participation, a grant of US$2 million was being provided under the World Bank's Governance Partnership Facility (GPF) to support a wide range of activities to be championed and implemented by civil society and community-based organizations (World Bank 2011).

Despite their contributions in many African countries, CSOs and NGOs dedicated to fostering good governance in the natural resource sector face a number of challenges. For example, in several African countries implementing the EITI, local civil society groups remained too weak to fulfill their watchdog function. Many are either co-opted or marginalized by the government; others simply lacked the capacity to hold governments and business to account (Carbonnier et al. 2011: 252). This was very much evident in Cameroon, where very few NGO committee members and civil society groups had the necessary training to understand the content of EITI reports, which greatly weakened their capacity to act as a watchdog of the transparency process (Gauthier and Zeufack 2011: 60).

In Sierra Leone, the actual capacity of civil society to hold the government and powerful companies accountable for their actions through the use of information made available by transparent reporting remains unclear (Maconachie 2009: 77). Such a notion may be unrealistic and unachievable in a fragile state such as Sierra Leone. Not only was the capacity for promoting transparency and accountability low, but relatively weak civil society actors remained largely unable to monitor or challenge the power of well-established rent-seeking actors who are firmly established in the mining sector, in some cases since the discovery of diamonds in the 1930s. Moreover, historically, there was not a culture of transparency or accountability in the mining sector in Sierra Leone, and the concept of record keeping was still very alien.

It is for reasons like these that Kolstad and Wiig (2008) argue that the activities of CSOs are insufficient in themselves to achieve transparency and must be complemented by other types of policies. This is because transparency depends on the electorate's level of education, the capacity and the extent to which stakeholders have the power to hold government accountable, and the private and collective nature of the goods about which information was provided.

Capacity development in natural resource management

A final policy initiative to improve natural resource management is embodied in the promotion of capacity development programs. Capacity development is defined by the Capacity Development Strategic Framework (CDSF) as "a process of enabling individuals, groups, organisations, institutions and societies to sustainably define, articulate, engage and actualize their vision or developmental goals building on their own resources and learning in the context of a pan-African paradigm." In addition, the CDSF states that rather than focusing on selected components, unblocking capacity challenges must be based on interventions that take into account the whole organization and system to be effective and sustainable (AU/NEPAD Capacity Development Strategic Framework 2009: 3).[4]

Despite its importance to resource-rich countries, many African countries lag behind their counterparts in other parts of the world when it comes to having the necessary capacity that would ensure the efficient and effective management of natural resources. The African Union Commission et al. (2011: 28) noted that with the exception of South Africa, the mineral sector in Africa generated little new knowledge in terms of mining-related products, processes, technologies, and services. Generally, indicators for the capacity for knowledge generation and innovation were the availability of scientists and engineers, quality of scientific research institutions, university-industry research collaboration, company spending on research and development, and government procurement of advanced technology products. However, these were all generally much lower in Africa than those found in emerging economies elsewhere.

In this vein, Collier and Venables (2011: 2) note that resource extraction companies from the developed and emerging economies could be presumed to know more than African governments about the chances of finding valuable resources. This information and the power asymmetry between African countries and their counterparts in other parts of the world, which stems from their limited capacity, weaken African countries' ability to engage in proper negotiations with foreign businesses operating in the natural resource sector. In fact, institutions that supported mineral development in Africa were generally weak. These weaknesses extend to training and education institutions; government departments responsible for formulating policies, laws, and regulations related to mining; and institutions in charge of negotiating mineral development agreements and monitoring and regulating the exploitation of mineral resources (African Union Commission et al. 2011: 18). As a result, bilateral negotiations between an African government and a prospecting company are likely to favor the company.

In Chad, for example, the public sector did not have the material and human resources to effectively manage the country's rapidly increasing revenue. Civil servants lacked the training to master the technical aspects of monitoring oil production and determining revenue. Those who demonstrated capacity were lured away to the private sector or became part of the increasing brain drain to the West that became common among Africa's technical class (Bryan and Hofmann 2007). The lack of basic and specialized training on contracts, negotiations, and dispute settlement, among others, meant that many African countries were at a disadvantage when negotiating technical and complex issues in the extractive sector with foreign companies. African governments' ability to negotiate favorable agreements with foreign companies will largely be determined by the availability of human capital to study various documents and advocate for their interests.

The way forward: Policy recommendations for promoting efficient natural resource management

The discovery of natural resources may not necessarily help improve the socioeconomic development of a country; it can, however, contribute to violence, social tension, insurgency, and instability. It is in this regard that governments, civil society, and other stakeholders have actively pursued various initiatives to improve natural resource management in Africa. Despite their laudable efforts, the initiatives to ensure that natural resources became a blessing as opposed to a curse have been bedeviled by a number of challenges. These experiences demonstrate the path that should be charted to ensure that natural resources are managed efficiently and thereby promote socioeconomic development in resource-rich African countries.

One of the first of the steps to promote the efficient and effective management of natural resources and thus stem the tide of the resource curse is to pursue policies that ensure that benefits are equitably distributed. As the two cases of transboundary water cooperation and governance discussed above showed, the allocation of water and related natural resources is something that must be mutually beneficial and agreed upon by all riparians involved.

As Chikozho (2012: 179–186) argues, the higher the net benefit perceived, the higher the likelihood of cooperation among riparians. Moreover, the possibility that a nation could improve its well-being by avoiding conflict and coordinating its actions with other riparians acts as a strong incentive to create institutions that could sustain basic cooperation. In addition, working toward effective governance requires an enabling environment and institutional structures that enhance stakeholder cooperation (Chikozho 2012: 157).

To this end, systematic efforts should be concentrated on policy harmonization and the implementation of mechanisms for conflict avoidance, conflict management, and interstate cooperation. Incentives for stronger interstate cooperation should be found or created, perhaps through cost- and benefit-sharing agreements that could be created and sustained by transboundary river basin organizations, which provide neutral platforms for dialogue in the basins and bring seemingly disparate riparian nations to the negotiation table (Chikozho 2012).

In particular, since the inequitable distribution of benefits contributed to tensions and violence in countries such as Nigeria, Liberia, and Sierra Leone, successful resource management would be dependent on how benefits accrue to citizens and the communities where the natural resources are located. Ensuring that some of the wealth from natural resources is repatriated to the communities in the form of productive community projects could lay the foundation for effective resource governance. For example, providing public goods, infrastructural development projects, and income-generating investments to help protect the communities in the Niger Delta region against some of their social and economic challenges and problems would be crucial to resolving the violent conflict (Ahonsi 2011). Therefore, it is necessary to formulate a strategy by which both national and local economies can benefit from a more effective and sustainable form of exploitation of natural resources. The essence of such a strategy is that it should ultimately be in the interests of local communities to operate both efficiently and transparently in the natural resource sector. But this will occur only if a fair proportion of the benefits are returned to local people and if, both nationally and locally, there are tangible gains from natural resources (Maconachie and Binns 2007).

Botswana, for example, not only has established a rainy-day fund to hedge against future unexpected economic shocks, but also has used its diamond resources to invest heavily in the health and education sectors. Public health facilities are distributed across the country and are now close to providing antiretroviral therapy to the estimated 100,000 citizens infected with HIV and in need of treatment for AIDS. Five thousand miles of paved and well-maintained roads allow travelers to move effortlessly throughout the large country. Free public education is provided to all children up to age 13, after which only a small tuition fee is required. The government finances almost the entire cost of the education of 12,000 students to study at the University of Botswana, established in 1982. Another 7,000 young persons study overseas on full government scholarships. High-quality game parks and wildlife management areas cover one-third of the country and generate substantial tourism revenue (McFerson 2009a: 1543). Moreover, by collecting and efficiently managing the revenue, and investing for development, Botswana has demonstrated how good natural resource management and governance can address social inequities and improve the lives of citizens.

Good governance affords a second means by which effective and efficient natural resource management could come about in Africa. A system of governance based on the rule of law, transparency, and accountability could address some of the challenges associated with resource mismanagement in African countries (Collier 2010; Arthur 2012: 122). As indicated by the African Union Commission et al. (2011: 24), transparent and participatory governance processes, at all levels, can assist mineral-rich countries in attaining sustainable economic growth and socioeconomic development. This is because public participation gave legitimacy to a project and reduced the costs emanating from the social tensions that could result from an externally imposed project.

Moreover, the existence of transparency and accountability ensured that political elites did not take advantage of their position to enrich themselves and family members through corrupt subcontracts and oil-trading practices while the rest of society was marginalized and impoverished. It also ensured that revenue from natural resources was not controlled by ruling cliques and government officials who entered into convenient negotiations with companies operating in the resource sector (Arthur 2012: 124). The promotion of good governance helps not only to mobilize and enhance the performance of the untapped potential of many African countries, but, more important, to achieve the CDSF's goal of leadership transformation and utilizing African skills, potentials, and resources for development. It is in this regard that resource-rich countries that were already engaged on the road to promoting good governance represented a vital opportunity to break the resource curse. Accordingly, these countries merited energetic international support so that they could reset their institutional incentives away from the unaccountable norms they have inherited (Siegle 2005: 50).

As stated earlier, nowhere has a positive strategy to resource management been more evident than in Botswana, where good governance—the establishment of strong and transparent governance structures, anticorruption systems, and integrity of public institutions—has contributed to its economic success. In addition to establishing a well-functioning judicial system that respected property rights and the rule of law, it adopted a consultative process in its decision-making process that involved traditional authorities (Hillbon 2008; McFerson 2009a; Taylor 2012). Aside from that, there exist independent, professional, and competent institutions, and there are clear rules as to how discontents ought to be monitored and dealt with in a manner that ensures that all have some share in the diamond resources (Gyampo 2011: 64).

Furthermore, given the capacity constraints that many African countries face and the critical technical challenge they present to the African natural resource sector's competitiveness (African Union Commission et al. 2011: 28), providing training and other capacity development programs for state and non-state actors in resource-rich African countries could complement the good governance process and thus be a key part of efforts to ensure efficient and effective resource management.

When a country has a paucity of the trained and skilled human resources required for tackling its particular problems, it struggles to resolve its development challenges (Ahonsi 2011: 36). For example, Nigeria's ability to resolve the violent conflict in the Niger Delta would require considerable investment in human capacity and institutional strengthening (Ahonsi 2011: 41). By developing capacity, being well informed, using knowledge and analysis, and striving to achieve what the New Partnership for Africa's Development (NEPAD's) CDSF calls knowledge-based and innovation-driven decision and development processes, African countries could improve the individual and institutional performance. For this reason, one cannot underestimate the important role of universities and other tertiary institutions in efforts to achieve the goals and objectives of improving the capacity African countries. Increasing funding to universities and research institutions and providing training and capacity-building programs in negotiations, agreements, and dispute settlements, among other areas of expertise, would allow African countries develop the practical knowledge and skills that could ultimately help them with natural resource manage-ment (Arthur 2004). There was thus the need for African countries to develop initiatives and policies that would address the lack of local capacity, and thereby offer opportunities for local expertise to actively participate in and manage the natural resource sector.

To realize this, it is important to increase the number of graduates with training in natural resource management and foster an institu-tional and supporting environment that enables those in the sector to thrive in their work. Moreover, capacity development programs could be created to build the capacity of research institutions and establish a positive relationship among the research institutions and workers in the resource sector through a process where trained technical extension specialists serve the resource sector. Support for resource management training at the tertiary level would help provide the necessary expertise and professionals working and involved in the natural resource sector.

To improve capacity development in the resource sector, there is the need for increased investment in improving the resources knowledge infrastructure. This would ensure that resource-rich African countries not only know the actual level and quality of their natural resource potential, but also the optimal strategy for determining and collecting adequate rent. Creating African capacity to sustain auditing, moni-toring, regulating, and improving resource exploitation regimes and developing resource sector linkages into the domestic economy should be another goal of African countries. This can be facilitated by ensuring that there is a skills transfer dimension in all contracted consultancies during lease/license negotiations as well as a targeted strategy around the development of such an ongoing resources governance capacity (African Union Commission et al. 2011).

In this vein, a welcome development in Ghana, for example, is the effort to build local capacity in the oil sector. According to *Daily Graphic* (2012), Tullow Oil plc, Africa's largest independent oil company, in part-nership with the British Council in Accra, launched the Tullow Group Scholarship Scheme to provide annual scholarships to 50 Ghanaians to pursue postgraduate programs abroad. The British Council man-ages candidate recruitment and selection according to criteria agreed with Tullow. A pilot phase of the scheme, which began in September 2011, provided funds to 24 Ghanaians from the public sector to pursue master's degree programs at leading universities in the United King-dom. The scholarship scheme is intended to enable local people to participate in the oil and gas industry and other sectors that promote economic diversification. It is anticipated that the scheme would address both existing industry skills gaps and national capacity development requirements consistent with Tullow's aim of supporting long-term socioeconomic growth in the company's operational areas.

Finally, African countries and stakeholders could promote efficient management and governance of natural resources by coordinating and integrating planning agencies operating across various sectors. The AU/NEPAD Capacity Development Strategic Framework (2009) notes that improving capacity should involve employing appropriate African technology that supports the continent's development challenges, participating in and owning the decision-making process rather than having it imposed by international actors, mobilizing domestic sources of revenue, and efficiently utilizing financial resources. Additional recommendations include monitoring and evaluating capacity development programs' performance and degree of success in meeting their stated goals and objectives, creating an environment that stems brain drain, and creating a system for ensuring that the best brains share their experiences in their countries (AU/NEPAD Capacity Development Strategic Framework 2009).

Concluding remarks

Many African countries are undoubtedly endowed with natural resources. Despite their potential to improve the socioeconomic conditions of their citizens, natural resources have not always had a positive impact on many countries. While expected to assist with the socioeconomic development, natural resource-rich countries have instead been bedeviled by conflicts, civil wars, and the Dutch disease. It is in this regard that adopting the right measures and policies would be crucial to ensuring that natural resources become a blessing as opposed to a curse in Africa. While initiatives such as the KPCS and EITI are a positive first step, these initiatives, as noted, face a number of challenges, with the most significant being its voluntary nature. It is in this regard that the chapter argues that capacity development programs, and the promotion of good governance, as seen in a country like Botswana, represent the most significant step toward the efficient management of natural resources among African countries. Good and better governance does not easily come about, but has to be effected through incentives (de Soysa 2011), ensuring that the equitable distribution of benefits from natural resources will be a step in the right direction in terms of ensuring that good governance is attained in resource-rich African countries.

Notes

1. While there is a lack of consensus on what constitutes a failed state, for Rotberg (2003: 5), "failed states are tense, deeply conflicted, dangerous, and bitterly contested by warring factions. In most failed states, government troops battle armed revolts led by one or more warring factions." Also, not only are basic functions of the state no longer performed, but also there is increased violence and the state is no longer able to deliver positive political goods to their people. The government loses legitimacy, and in the eyes and hearts of a growing plurality of its citizens, the nation state itself becomes illegitimate.
2. According to Basedau and Lay (2009: 761), the concept of the rentier state was developed with regard to the Middle East oil producing states such as Iran and Gulf Monarchies and argues that the main function of the state in rentier economies is to distribute rent. The rents provide ruling elites with vital revenue and other resources through which to address and offset any potential pressures for violence and instability.
3. The Orange River basin is one of the international watercourse systems in the SADC and of strategic importance to South Africa, Lesotho, Namibia, and Botswana (Heyns et al. 2008).
4. The AU/NEPAD Capacity Development Strategic Framework (CDSF) seeks to guide capacity development and also address significant knowledge-based capacities toward addressing Africa's development challenges. The cornerstones of the CDSF are leadership transformation, citizen participation, utilizing African potentials, skills, and resources for development, knowledge-based and innovative-driven processes, and capacity of capacity builders.

References

African Union Commission, African Development Bank & United Nations Economic Commission for Africa. (2011). "Building a Sustainable Future for Africa's Extractive Industry: From Vision to Plan: Action Plan for Implementing the AMV," Available at: http://www.africaminingvision.org/amv_resources/AMV/Action%20Plan%20Final%20Version%20Jan%202012.pdf. Accessed August 24, 2012.

Ahonsi, B. (2011). "Capacity and Governance Deficits in the Response to the Niger Delta Crisis," in C. Obi and S. Rustad (eds.), *Oil and Insurgency in the Niger Delta: Managing the Complex Politics of Petroviolence*, London: Zed Books, pp. 28–41.

Alao, A. (2007). *Natural Resources and Conflict in Africa: The Tragedy of Endowment*, Rochester, NY: University of Rochester Press.

Arthur, P. (2004). "The Multilateral Trading System, Economic Development and Poverty Alleviation in Africa," *Canadian Journal of Development Studies*, 25(3): 429–444.

Arthur, P. (2010). "Democratic Consolidation in Ghana: The Role and Contribution of State Institutions, Civil Society and the Media," *Commonwealth and Comparative Politics*, 48(2): 203–226.

Arthur, P. (2012). "Averting the Resource Curse in Ghana: Assessing the Options," in L. Swatuk and M. Schnurr (eds.), *Natural Resources and Social Conflict: Towards Critical Environmental Security*, London: Palgrave Macmillan, pp. 108–127.

AU/NEPAD Capacity Development Strategic Framework. (2009). Available from: http://www.oecd.org/development/governanceanddevelopment/43508787.pdf. Accessed October 20, 2012.

Basedau, M. and Lay, J. (2009). "Resource Curse or Rentier Peace? The Ambiguous Effects of Oil Wealth and Oil Dependence on Violent Conflict," *Journal of Peace Research*, 46(6): 757–776.

Bryan, S. and Hofmann, B. (2007). *Transparency and Accountability in Africa's Extractive Industries: The Role of the Legislature*, Available from: www.accessdemocracy.org/files/2191_extractive_080807.pdf. Accessed September 2, 2012.

Carbonnier, G., Brugger, F., and Krause, J. (2011). "Global and Local Policy Responses to the Resource Trap," *Global Governance*, 17: 247–264.

Chikozho, C. (2012). "Towards Best-Practice in Transboundary Water Governance in Africa: Exploring the Policy and Institutional Dimensions of Conflict and Cooperation over Water," in K. Hanson, G. Kararach, and T.M. Shaw (eds.), *Rethinking Development Challenges for Public Policy*, London: Palgrave Macmillan, pp. 155–200.

Collier, P. (2007). *The Bottom Billion, Why the Poorest Countries Are Failing and What Can Be Done about It*, New York: Oxford University Press.

Collier, P. (2010). *The Plundered Planet: Why We Must—And How We Can—Manage Nature for Global Prosperity*, New York: Oxford University Press.

Collier, P. and Venables, A. (2011). "Key Decisions for Resource Management: Principles and Practice," in P. Collier and A. Venables (eds.), *Plundered Nations? Successes and Failures in Natural Resource Extraction*, London: Palgrave Macmillan, pp. 1–26.

Daily Graphic. (2012). "Tullow to Support Training of Graduates in Oil and Gas," Available from http://edition.myjoyonline.com/pages/education/201201/80499.php. Accessed September 29, 2012.

Daniele, V. (2011). "Natural Resources and the 'Quality' of Economic Development," *Journal of Development Studies*, 47(4): 545–573.

de Renzio, P., Gomez, P., and Sheppard, J. (2005). "Budget Transparency and Development in Resource-Dependent Countries," *International Social Science Journal* (Supplement 1), 57: 57–69.

de Soysa, I. (2011). "The Natural Resource Curse and State Failure: A Comparative View of Sub-Saharan Africa," in M. Roll and S. Sperling (eds.), *Fuelling the World—Failing the Region? Oil Governance and Development in Africa's Gulf of Guinea*, pp. 34–52, Available from http://library.fes.de/pdf-files/bueros/nigeria/08607.pdf. Accessed October 17, 2012.

EITI International Secretariat. (2010). *Impact of EITI in Africa: Stories from the Ground*, Available at: http://eiti.org/files/EITI%20Impact%20in%20Africa.pdf. Accessed August 30, 2012.

Fearon, J.D. (2005). "Primary Commodity Exports and Civil War," *Journal of Conflict Resolution*, 49(4): 483–507.

Gary, I. (2009). *Ghana's Big Test: Oil's Challenge to Democratic Development*, Oxfam America and the Integrated Social Development Centre. Available at: http://www.publishwhatyoupay.org/sites/pwypdev.gn.apc.org/files/Ghanas-Big-Test%20OxfamISODEC.pdf. Accessed August 30, 2012.

Gauthier, B. and Zeufack, A. (2011). "Governance and Oil Revenues in Cameroon," in P. Collier and A. Venables (eds.), *Plundered Nations? Successes and Failures in Natural Resource Extraction*, London: Palgrave Macmillan, pp. 27–78.

Gonzalez-Vicente, R. (2011). "China's Engagement in South America and Africa's Extractive Sectors: New Perspectives for Resource Curse Theories," *Pacific Review*, 24(1): 65–87.

Grant, J.A. and Taylor, I. (2004). "Global Governance and Conflict Diamonds: The Kimberley Process and Quest for Clean Gems," *Round Table*, 93(375): 385–401.

Gyampo, R.E.V. (2011). "Saving Ghana from Its Oil: A Critical Assessment of Preparations So Far Made," *Africa Today*, 57(4): 48–69.

Heyns, P., Marian, P., and Turton, A. (2008). "Transboundary Water Resource Management in Southern Africa: Meeting the Challenge of Joint Planning and Management in the Orange River Basin," *International Journal of Water Resources Development*, 24(3): 371–383.

Hillbon, E. (2008). "Diamonds or Development? A Structural Assessment of Botswana's Forty Years of Success," *Journal of Modern African Studies*, 46(2): 191–214.

Humphreys, M., Sachs, J. and Stiglitz, J. (eds.) (2007). *Escaping the Resource Curse*, Irvington, NY: Columbia University Press.

Idemudia, U. (2009). "The Quest for the Effective Use of Natural Resource Revenue in Africa: Beyond Transparency and the Need for Cultural Compatibility in Nigeria," *Africa Today*, 56(2): 1–24.

Ite, U. (2004). "Multinationals and Corporate Social Responsibility in Developing Countries: A Case Study of Nigeria," *Corporate Social Responsibility and Environmental Management*, 11(1) (March): 1–11.

Kagwanja, P. (2007). "Calming the Waters: The East African Community and Conflict over the Nile Resources," *Journal of Eastern African Studies*, 1(3): 321–337.

Karl, T.L. (1997). *The Paradox of Plenty: Oil Booms and Petro States*, Berkeley: University of California Press.

Karl, T.L. (1999). "The Perils of the Petro-State: Reflections on the Paradox of Plenty," *Journal of International Affairs*, 53(1): 31–48.

Kolstad, I. and Wiig, A. (2008). "Is Transparency the Key to Reducing Corruption in Resource-Rich Countries," *World Development*, 37(3): 521–532.

Le Billon, P. (2008). "Diamonds Wars? Conflict Diamonds and Geographies of Resource Wars," *Annals of Association of American Geographers*, 98(2): 345–372.

Maconachie, R. (2009). "Diamonds, Governance and 'Local' Development in Post-Conflict Sierra Leone: Lessons for Artisanal and Small-Scale Mining in Sub-Saharan Africa?" *Resources Policy*, 34(1/2): 71–79.

Maconachie, R. and Binns, T. (2007). "Beyond the Resource Curse? Diamond Mining, Development and Post-Conflict Reconstruction in Sierra Leone," *Resources Policy*, 32(3): 104–115.

McFerson, H. (2009a). "Governance and Hyper-Corruption in Resource-Rich African Countries," *Third World Quarterly*, 30(8): 1529–1548.

McFerson, H. (2009b). "Measuring Governance: By Attributes or Results?" *Journal of Developing Societies*, 25(2): 253–274.

McFerson, H. (2010). "Extractive Industries and African Democracy: Can the Resource Curse Be Exorcised?" *International Studies Perspectives*, 11(4): 335–353.

Mehlum, H., Moene, K., and Torvik, R. (2006). "Institutions and the Resource Curse," *Economic Journal*, 118(508): 1–20.

Mirumachin, N. and Van Wyk E. (2010). "Cooperation at Different Scales: Challenges for Local and International Water Resource Governance in South Africa," *Geographical Journal*, 176(1): 25–38.

Obi, C. (2010). "Oil as the 'Curse' of Conflict in Africa: Peering through the Smoke and Mirrors," *Review of African Political Economy*, 37(126): 483–495.

Obi, C. and Rustad, S.A. (2011). "Petro-Violence in the Niger Delta: The Complex Politics of an Insurgency," in C. Obi and S. Rustad (eds.), *Oil and Insurgency in the Niger Delta: Managing the Complex Politics of Petroviolence*, London: Zed Books, pp. 1–13.

Omorogbe, Y.O. (2006). "Alternative Regulation and Governance Reform in Resource Rich Developing Countries of Africa," in B. Barton, L. Barrera-Hernandez, A. Lucas, and A. Ronne (eds.), *Regulating Energy and Natural Resources*, Oxford: Oxford University Press, pp. 39–65.

Prempeh, K. and Kroon, C. (2012). "The Political Economy Analysis of the Oil and Gas Sector in Ghana: Summary of Issues for Star-Ghana," Available at: http://www.starghana.org/assets/STAR%20Ghana%20Recommendations%20 and%20Summary%20of%20Issues%20for%20Oil%20&%20Gas%20Call.pdf. Accessed August 31, 2012.

Rotberg, R. (2003). *State Failure and State Weakness in a Time of Terror*, Washington, DC: Brookings Institution Press.

Sachs, J. and Warner, A. (2001). "The Curse of Natural Resources," *European Economic Review*, 45(4–6): 827–838.

Siegle, J. (2005). "Governance Strategies to Remedy the Natural Resource Curse," *International Social Science Journal*, Supplement 1(57): 45–55.

Sovacool, B. (2010). "The Political Economy of Oil and Gas in Southeast Asia: Heading towards the Natural Resource Curse?" *Pacific Review*, 23(2): 225–259.

Swatuk, L. (2012). "Water and Security in Africa: State-Centric Narratives, Human Insecurities," in L. Swatuk and M. Schnurr (eds.), *Natural Resources and Social Conflict: Towards Critical Environmental Security*, London: Palgrave Macmillan, pp. 83–107.

Taylor, I. (2012). "Botswana as a Development-Oriented Gate-Keeping State: A Response," *African Affairs*, 111(444): 466–476.

Teshome, W. (2008). "Transboundary Water Cooperation in Africa: The Case of the Nile Basin Initiative," *Alternatives: Turkish Journal of International Relations*, 7(4): 34–43.

UNRISD (United Nations Research Institute for Social Development). (2007). Report of the UNRISD International Workshop, March 1–2, UNRISD Conference News.

Wantchekon, L. (2002). *Why Do Resource Abundant Countries Have Authoritarian Governments?* Available from: http://www.afea-jad.com/2002/Wantchekon3 .pdf. Accessed September 1, 2012.

Wexler, L. (2010). "Regulating Resource Curses: Institutional Design and Evolution of Blood Diamond Regimes," *Cardozo Law Review*, 31(5): 1717–1780.

World Bank. (2011). "Building Capacity to Manage Ghana's Oil—World Bank Assists with US$38 Million," Press Release No. 2011/272/AFR, Available at: http://web.worldbank.org/WBSITE/EXTERNAL/COUNTRIES/AFRICAEXT/ GHANAEXTN/0,,contentMDK:22794423~menuPK:351972~pagePK:28650 66~piPK:2865079~theSitePK:351952,00.html. Accessed September 4, 2012.

Wright, C. (2004). "Tackling Conflict Diamonds: The Kimberley Process Certification Scheme," *International Peacekeeping*, 11(4): 697–708.

4
Criminality in the Natural Resource Management Value/Supply Chain

Olawale Ismail and Jide Martyns Okeke

Introduction

Since the beginning of the 1990s, African development and political security research and policy discourse has been dominated by issues of natural resource endowments, extraction, and receipts and the range of actors and processes involved, often unpacked as the natural resource "value chain." The discourse throws up issues of natural resource governance—the framework of rules, institutions, and practices regulating the natural resource value chain and the extent to which key principles of transparency, openness, accountability, fairness, and environmental sustainability are observed in the extraction, movement, and receipts from natural resources. The plethora of research on natural resource in Africa focuses on its connections to armed conflicts and peace building. Phrases such as natural resource "curse" or conflict, and "paradox of plenty" have emerged to underscore the negative impacts of natural resource issues on state and society in Africa. Criminality in the natural resource value chain is generally, yet restrictively, subsumed within the discourse of its nexus with armed conflicts.

However, criminality in the natural resource value chain is more complex and multilayered, and transcends conventional categorizations, as it often entails official (state-led) and unofficial (unrecorded) dimensions; first, it involves actors and processes linked to practices in the public, private, and civil society spheres; is hierarchically syndicated with local, national, regional, and international actors; and transcends the orthodox conflict-and-peace divide, as it takes place in stable and conflict-affected countries. Second, existing mechanisms for addressing this issue are limited and often focused on law enforcement constructs rather than alongside the sociocultural, political, environmental, and economic undercurrents of criminality in the natural resource value chain. Third, acute capacity deficits in enforcement mechanisms arise from policy gaps, the pressures of globalization, insecurity, institutional decay, and inefficiencies linked to pervasive corruption. The policy gaps relate to unresolved contestations between extant laws and normative, informal practices related to the ownership, extraction, and management of receipts from natural resources. All this underscores the need for a holistic approach to understanding and designing policy interventions in combating criminality in the natural resource value chain in Africa.

This chapter contextualizes natural resources to include "all non-artificial products situated on or beneath the soil, which can be extracted, harvested, or used, and whose extraction, harvest, or usage generates income or serves other functional purposes in benefiting mankind" (Alao 2007: 16). A substance is defined as a natural resource where and when it is naturally occurring and considered valuable by and in its unmodified form. Criminality relates to acts, actions, and inactions that, by commission or omission, cause harm and deprivation, endanger the commonwealth, and are prohibited by extant laws and social norms in the natural resource value chain. Criminality is a relative term, grounded in a social construct that reflects societal and governmental realities and prohibitions within a particular historical milieu.

The focus here is on timber or logging (renewable natural resource) and oil (nonrenewable natural resource). We emphasize renewable natural resources to be substances that can be replaced, replenished, and regenerated over time and nonrenewable natural resources to be substances that cannot be replenished or regenerated or replaced. Liberia is the case study for interrogating the key dynamic of criminality in timber resource management, while Nigeria is used for assessing criminality issues in oil. The two countries have dominated the discourse of criminality in the natural resource value chain in Africa; they have parallel experiences of armed insurgencies linked to natural resources, and are currently attempting to reform their natural resource sectors. When juxtaposed against countries such as Botswana, South Africa, Ghana, and Zambia, it is not impossible that the two case studies represent worst-case scenarios in terms of the depth of challenges confronting the (mis)management of natural resources in Africa. Still, Liberia and Nigeria provide invaluable empirical opportunities to see the nature and scope of criminal practices, policy responses, and potential and actual progress in improving the management of the natural resource value chain in Africa. Accordingly, the case studies are used to tease out the nature, scale, and dimensions of the problem; to provide an overview of the key local, national, and international actors involved in the criminal networks profiteering from the natural resource value chain; and to analyze and pinpoint gaps in policy responses.

Case study 1—Liberia: "Legitimizing" criminality through private use permits in logging?

Illicit timber trade is one of the major aspects of transnational organized environmental crimes. Timber trade (both legal and illegal) has historically been a critical part of international commerce. According to the Food and Agriculture Organization (FAO), the annual turnover of international trade in timber from the 1990s to 2009 was more than US$200 billion (cited in TRAFFIC 2012). These data do not capture the often clandestine and unofficial trade in illicit timber. For instance, in 2009, the global trade in illicit timber to the European Union (EU) from China and Southeast Asia alone was estimated at US$2.6 billion (UNODC 2010). Similar patterns in the export of illicit woods from African states can also be identified; however, it is more difficult to find reliable data due to the dearth of reliable research on this issue. The problem is complicated by the peculiarities of illicit timber trade, especially the fact that some of this trading occurs within the context of heightened insecurity caused by civil wars or the activities of armed criminal groups.

Notwithstanding, it is possible to identify some common patterns and actors involved in illegal logging. The trade is often facilitated by the use of fraudulent paperwork that ensures that protected woods, for instance, can be transmuted into a more mundane variety (UNODC 2010: 10). By so doing, the restrictions or prohibitions of trade in specific protected timbers are bypassed through fraudulent documentation. This does not necessarily mean that such documentation is forged or that these activities operate within a largely dominant unofficial network. On the contrary, illegal logging networks (warlords, gangs, multinational companies) tend to work closely with some government officials in facilitating this trade. Accordingly, illicit logging can be initiated and sustained through corrupt practices (government clearance of protected woods through official certification), the issuance of questionable land permits to national and multinational stakeholders, or direct stealing of woods (sometimes with or without government knowledge or endorsement). Some of these features are discernible in the pattern of illicit timber trade in Liberia.

Liberia has ostensibly emerged as one of the success stories of post-conflict transition in Africa. Since the election of Ellen Johnson Sirleaf, Africa's first female president, there is general optimism about political stability and economic development in Liberia. The World Bank (2012) submits that Liberia's economy grew by 4.6 percent and 5.6 percent in 2009 and 2010, respectively. A substantial part of this economic performance has been attributed to the increase in agricultural outputs (especially rubber and forestry).

Despite this significant progress, it is important to examine criminality issues in Liberia's timber industry with a view to analyzing the undercurrents of illegal logging. While there has been much progress in addressing "conflict logging," which characterized wartime Liberia, the current regime against illegal logging in Liberia appears to be reinforcing and legitimizing illegal logging and reduces the prospect of sustainability.

Nature and evolution of illegal logging

The timber value chain in postwar Liberia can be best understood in terms of its structure and actors. Structurally, the government remains the primary authority in granting land licenses for timber exploitation and its eventual delivery to final customers, but there are official provisions for community forests—areas protected from commercial logging activities and reserved for the sociocultural and economic well-being of host communities. The Forestry Development Authority (FDA) is the lead government agency responsible for issuing licenses and overall regulation of the timber industry. The timber companies, mainly owned by foreign entrepreneurs and countries, directly exploit timber and other forest resources in exchange for rents and commissions to the Liberian state (government). There are also local communities and their inhabitants in and around the forest belts who operate on the margins of the formal process as laborers, traditional owners of farmlands and forestlands, and providers of support services to timber companies.

Generally, criminality in Liberia's timber industry takes place at almost all layers and involves almost all the actors in the value chain. Criminal practices are perpetrated at the level of issuing licenses (often political decisions), performing regulatory and oversight functions by the FDA, monitoring and reporting accurate data, and remitting appropriate rents and taxes by logging companies; informal (unapproved) logging practices by companies and private individuals further complicate the landscape.

Illicit logging in Liberia has a long a history; however, this analysis focuses on the post-Cold War period. The post-1989 illicit logging trade in Liberia has two main phases. The first falls within the period of protracted civil wars in Liberia that date from 1989 to 2003. There was a precarious interlude in the war efforts following the 1997 election of Charles Taylor, a former leader of the rebel National Patriotic Front of Liberia (NPFL), which ousted and killed President Samuel Doe. However, there was a relapse into armed insurgencies during Charles Taylor's reign as new rebel groups launched attacks from neighboring Guinea. Numerous studies establish that until 2003, when Taylor resigned from office, he used revenue from illegal logging to enrich himself and finance his war efforts (Reno 2000; Richards 2001; Beevers 2012).

A useful way of understanding illicit conflict logging in Liberia is through a "business and diplomacy" model. This posits that the collapse of state institutions, which was preceded and exacerbated by Liberia's civil war, inspired "domestic and international interests to scramble for political advantages and control over resources" (Ellis 1999: 164). The domestic interests were led by Charles Taylor, his relatives, and senior military commanders (ICG 2003). A 2002 Global Witness report documents the involvement of Charles Taylor, Jr., and Demetrius Taylor (Taylor's brother), who served respectively as Managing Director and Secretary of the FDA, in illegal logging. There were also long-standing international commercial interests in Liberia's timber, which contributed to the flourishing of the illicit logging. This included links between multinational companies (and other countries) with the Liberian logging industry. For example, Abbas Fawaz, one of Charles Taylor's leading financial supporters, benefited significantly from illicit timber trade through his logging firm, United Logging Company. In addition, other multinational companies from France, the Netherlands, and China were also beneficiaries of this illicit trade. In 2001, for example, France imported 98,700 cubic meters (m^3) of Liberian logs estimated at US$13.2 million. The Netherlands—through Wijima, one of the largest Dutch hardwood importers—was also involved in this trade. China remained the largest importer of Liberian timber into Asia (Global Witness 2012a). These multinational "logging companies made significant off-budget payments to Taylor and were actively involved in illegal arms exports" (Global Witness 2012a: 1).

The illicit trade was carried out via sea and air transport, and via land (using neighboring countries, especially Côte d'Ivoire), with Liberian timber exchanged for cash and/or arms and ammunition. Before the outbreak of the civil war in 1989, the illicit timber trade had flourished in the context of the Liberian state's diminished capacity to monitor, regulate, and counteract illegal commerce of its timber. In other words, illicit timber trade significantly contributed to the destruction of Liberia's economy and the cycle of violence during the 1990s (Ellis 1999).

Since the end of the war and the election of Ellen Johnson Sirleaf, there have been attempts to reform and improve the regulation of the timber industry. The pillar of reform was the establishment of the private use permits (PUP) under the 2006 National Forestry Reform Law (NFRL), intended to regulate and govern the logging trade, especially the relations between private landowners and government-approved companies. Yet, it seems that this process has been less sensitive to environmental sustainability (destroying rainforest), continues to be characterized by corrupt and fraudulent practices, and harbors risks of community-based conflicts.

The logging syndicate: Warlords, government officials, and foreigners

When Charles Taylor and other warlords engaged in looting and illegal exploitation of timber during Liberia's civil war, they were able to amass revenue through informal networks with Western and non-Western companies and states. For example, it was reported that France and China imported an estimated 71 percent of Liberia's timber in 2000, and the United Kingdom, Italy, Denmark, Germany, and other countries were also part of the conflict-logging global trade network (Global Witness 2001).

There are no established links between President Sirleaf and illicit logging in post-conflict Liberia; however, concerns have been expressed about the complicity of some government officials in facilitating the access of multinational logging firms to land deeds through corrupt practices. Multinational companies remain the most influential players in Liberia's timber industry. They have often sought to maximize profit through nonpayment of taxes, illicit acquisition of land deeds, and undue influence on government decisions, especially when they are unfavorable to their businesses. The companies that maintain dominant positions in Liberia's timber industry include Atlantic Resources, Alpha Logging, and Samling. These three companies hold a total of more than 20 percent of Liberia's land area (Global Witness 2012b: 4). In 2008, the Save My Future Foundation issued a report alleging that Tobga Timber Company is responsible for felling, sawing, and exporting timber illegally from Maryland County, which had implications for rebellion in Côte d'Ivoire. Similarly, Atlantic Resources, which is linked to the Malaysian logging company Samling, has been extensively implicated in logging practices that are exploitative of community-owned rainforest.

The scale of logging permits awarded to multinational companies suggests that they remain important actors in the timber sector in Liberia. The logging permits are especially questionable on account of reduced consideration for community forests and sustainability (as concerns raised by civil society groups). The ecosystem depletion caused by excessive felling of trees may degenerate into conditions that breed conflicts in the future. Unfortunately, despite the reform put in place by the Liberian authorities, the illicit pattern of logging continues unabated. In fact, it seems that the legislations put in place by the government provides some legitimacy for various actors to continue the illicit practices.

Policy responses: From a sanction regime to a PUPs regime

During the Liberian civil war, the United Nations imposed wide range of sanctions against Charles Taylor's regime. However, it was not until 2003 that it decided to recognize the connection between timber trade and the sustained conflict in Liberia. Specifically, UN Security Council Resolution 1478 imposed a ten-month ban on imports of Liberian timber by UN member states.[1] This sanction was lifted in 2006 guided by UNSCR 1689 (UNSC 2006), but there have been major efforts in reforming the timber sector, especially in promoting sustainability and preventing illicit trading.

Major trading partners like the United States have offered development assistance to Liberia to promote sustainability in the timber value chain. For example, in September 2012, it was reported that the United States provided US$30 million to support communities in the management of forest resources. In addition, the European Union (on behalf of its member states) signed the Voluntary Partnership Agreement (VPA) with Liberia as the legal framework to set minimum standards on the felling of trees (FLEGT 2012). The NGO Coalition for Liberia worked closely with local communities on the signing, one example of the manner in which civil society organizations (CSOs) have contributed to and have remained important actors in the formulation, implementation, and monitoring of various laws guiding legal timber trade.

The Liberia Governance and Economic Management Assistance Program (GEMAP) was launched in 2005 as a framework for a post-conflict reconstruction program aimed at building formal institutions to curb massive corruption and prevent loss of government revenue, especially through illegal procurements and practices. The aim of GEMAP is to create and institutionalize effective financial and asset management policies and procedures, contain corruption, and improve overall economic governance. Specifically, the GEMAP process sets out six main objectives: to secure Liberia's revenue base; ensure improved budgeting and expenditure management; improve procurement practices and grants of natural resources concessions; establish effective processes to control corruption; support central institutions of government; and foster crosscutting capacity building (Cohen et al. 2010). This ambitious blueprint has been the framework for Liberia's postwar economy recovery, especially its natural resource governance architecture.

In the area of forestry, the GEMAP seeks to build the FDA's capacity through staff training, financial management systems, and periodic performance (financial and procedural) audits. GEMAP's notable achievements in relation to forestry management include standardization of all payment and revenue collection systems, improved documentation and monitoring through staff capacity training, establishment of safeguards against illegal and unqualified logging concessions, and detection of irregularities in the logging sector to safeguard against loss of official revenue (http://www.gemap-liberia.org/about_gemap/FDA.html).

The overarching national legislation governing the timber sector is the NFRL, adopted in 2006 as part of the system-wide reform to ensure sustainability and allow for the legal trading of timber. This reform agenda provides a framework for mediating disputes between landowners and government-approved companies in the timber trade. An important provision within the NFRL is the establishment of the private use permits (PUPs). The PUPs allows for the commercial use of forest resources on private land. Under the PUP framework, the Liberian government is expected to grant permits to private landowners that will allow them trade in timber and other forest resources with government-approved companies. Since its adoption, the government has awarded about 65 PUPs that are estimated to cover approximately one-third of Liberia's forest. The PUPs were established to prevent illegal trade in timber, minimize disputes between landowners and logging firms, and meet the increasing demand for Liberia's timber. Other benefits of the PUPs as originally conceived are to generate revenue through the payment of rents by PUP operators to the government; enhance oversight and regulation over the timber sector; attract foreign direct investments

(FDIs); and improve economic benefits to landowners at the community level through fair trade by government-approved companies.

Unfortunately, the PUPs may not have delivered the intended benefits, as the program seems to be entrenching illegality, unaccountability, and economic stagnation in Liberia vis-à-vis the timber sector. A 2012 study by Global Witness noted the following problems associated with the implementation of the PUPs:

1. There are very few restrictions on PUPs in Liberia's law, which allows for the possibility of widening the commercial exploitation of forest resources. This raises questions about the sustainability rationale behind the formulation of the NFRL and the PUPs in particular.
2. The government has been unable to document receipt of substantial revenue from the PUPs. This is because landowners and companies are able to agree on rates of compensation that will allow them to pay either little or no taxes to the government.
3. Communities are very vulnerable to the consequences of excessive exploitation of their forests. Hence, there is a high risk of erosion, deforestation, and land degradation. This is compounded by the fact that there have been instances of falsified land deeds and other corrupt practices (Global Witness 2012b: 3).
4. Finally, PUPs seem to be legitimizing the possibility of community exclusion and widespread exploitation of forest resources that are supposed to be preserved for communities. Under the NFRL, and specifically the 2009 Community Rights Law, the community forests provision makes certain lands (forests) ineligible for PUPs. Regrettably, some of the PUPs awarded violated this provision.

The inadequacies of the regulatory regimes have led to protests by CSOs and to temporary moratoriums on PUPs. For example, in response to those protests, the government issued two moratoriums (in February and August 2012) on the award of PUPs to landowners to reverse the negative trends of the PUPs. Even though much progress has been made in the shift away from conflict logging, aspects of the current government regulations inadvertently perpetuate the illegal pattern of logging in Liberia.

Post-conflict Liberia has made some progress in attempting to govern its resources as discernible in its timber sector. This is noticeable in the reform of policy formulation, institution building, and operational responses in curbing corruption and insecurity, and collaborating with international actors as a disincentive for reducing transnational illicit timber trade. There are, however, considerable capacitation gaps in the current regime governing natural resources in general and the timber trade in particular, both at the policy (especially the current PUP regime) and at the operational levels.

Case Study 2—Nigeria: Illegal bunkering in the oil sector

In September 2012, the Nigerian finance minister raised national and international alarm that the trade in stolen crude had reached an all-time high, reducing official exports by 17 percent (the equivalent of 400,000 barrels per day) and amounting to a loss of N1.2 billion (approximately US$7,667,000) in official revenue on a monthly basis (Globserver 2012). In November 2009, the ship *MT African Prince* was intercepted some 15 nautical miles off Ghana's territorial waters with 5,200 tons of crude confirmed to have been stolen from Nigeria (*All Africa News*, November 29, 2009). In October 2003, another ship, *MT African Pride* and its ten Russian crew members were arrested by the Nigerian navy for carrying some 6,500 metric tons of stolen crude oil from southern Nigeria. Some months thereafter, the ship and crew disappeared from official custody under mysterious circumstances that prompted an official investigation and the eventual court-martialing and dismissal of two rear admirals in January 2005.

These anecdotes illustrate different yet connected aspects, the cost, scales, actors, and networks involved in illegal bunkering in the value chain of Nigeria's oil sector. They detail the role of internal (Nigerian) and external actors in a well-organized syndicate, and the landscape of oil bunkering in Nigeria. Although the focus is illegal bunkering in the context of the upstream (extractive) sector, the landscape of criminality in Nigeria's oil value chain is wider as it includes corrupt practices in the management of oil receipts, and award of contracts and oil exploration licenses.

Bunkering generally refers to the licensed movement of condensate (refined and unrefined oil) from one point to another using water transport. Domestically, it is for easing the movement of oil among riverine areas. Internationally, it is a key element in merchant trade and the international energy market. Hence, there are legal and illegal forms of bunkering. Over the past decade, illegal oil bunkering has become more pronounced, intensified, and syndicated with the active collusion of security personnel, armed groups, vessel owners, riverine communities, and highly placed government officials in Nigeria. While the phenomenon predates 1999, it has taken a new dynamic ever since, becoming intertwined with issues of insecurity and the tensed tripartite relationship involving host communities, the government, and multinational oil companies operating in the Niger Delta region.

Located in the southern region of Nigeria, the Niger Delta is a huge expanse of swampy and semi-swampy area with complex networks of creeks, swamps, and waterways. The epicenter of oil exploration activities in Nigeria, with a proven reserve of over 36 billion barrels of crude,[2] it suffers from serious environmental challenges, of which oil-related pollution and gas flaring are key elements. The region's inhabitants are generally poor, despite the huge financial investments in and revenue from oil and gas exploration activities. This has generated serious social agitation and grievance that has mutated into a mixture of insurgency and criminality.

The nature of illegal oil bunkering

The trade in stolen crude oil in Nigeria has official and unofficial dimensions. *Official illegal bunkering* primarily involves using existing oil-lifting licenses to illegally take and profit from crude oil—the lifting of excess crude beyond approved quantities. It refers to the use of unlicensed ships, barges, and vessels to transport illegal crude and the falsification of documents on the lifting, volume, cargo movement, and destination through the active collusion of officers of the Customs Service, the ship captain, and officials at loading terminals. To underscore the reality of official bunkering, the chief of Nigeria's Customs Service announced at a briefing for legislators in August 2008 that his department lacked the operational capacity to perform its statutory mandate of monitoring and documenting the amount of crude taken out of Nigeria on a daily basis. The customs chief noted that

> today our maritime unit has collapsed, we cannot afford to buy sea going vessels or boats. How do we tackle cases of bunkering?... We don't even know how many vessels come into this country... We are supposed to know the quantity of crude that goes out of this country, but as I am talking now, only very few companies come and call us to go and inspect the vessel, only few of them because we don't have the facilities to do it ourselves.
>
> (ThisDay 2008)

There are suggestions that illegal oil bunkering is hardly new in Nigeria: that it actually started in the mid- to late 1970s following a massive increase in oil output and revenue, and the expansion of maritime business in Nigeria. The prevalent pattern up to the early 1990s was the cargo theft format perpetrated by serving and retired high-level military personnel who awarded oil-lifting contracts (licenses) to themselves or their cronies and transported stolen crude alongside official crude. According to one analyst, the cargo theft format of oil bunkering expanded massively circa 1996 during the regime of General Sani Abacha, when serving high-ranking naval chiefs smuggled stolen crude from the Niger Delta to neighboring countries in West Africa for sale. It is estimated that in the first six months of 1996, the three naval chiefs smuggled over 202,130 metric tons of crude out of Nigeria (Busch 2005). Since 1999, it is reported that key figures in the ruling People's Democratic Party (PDP), a group dominated by retired military chiefs, have actively encouraged and participated in unofficial oil bunkering, parallel to their collection of kickbacks for contracts in the oil and gas sector.[3]

Unofficial illegal bunkering involves stealing crude oil through a network of connected strategies. There are two stages involved in this process. First is the sourcing or extraction of crude: this is done mostly through direct extraction by hot tapping—drilling or bridging a crude pipeline, often at night when a crude production line has been shut down owing to sabotage (pipeline vandalization) that ideally causes a drop in pressure registered at gauges at flow stations. During the shutdown, oil bunkerers install sophisticated taps from which they load stolen crude when the production line is restarted. In most cases, this is very difficult to detect as the supply is always constant, thereby preventing any fluctuation in pressure following the resumption of production at a flow station. Second is "wellhead" tapping—the invasion of an abandoned production wellhead to remove the safety valves, and the reinstalling of production enabling valves to restart the pumping of crude. The actual process of stealing often takes place deep inside creeks and swamps using plastic containers and small and medium-size barges as intermediate storage facilities for onward transfer to ocean-going oil vessels and tankers.

The process of selling stolen crude on international black markets involves a complex network of local and foreign individuals who tend to coordinate the process from extraction to transportation and discharging. There are reports that stolen crude is actually sold on the international black market at a notorious "Togo Triangle"—a patch of maritime area, often not policed, in and around Togo's territorial waters where money and stolen crude exchange hands (*Vanguard* 2009b). From arrests made and media reports, the dealers (sellers and buyers) of stolen crude from the Niger Delta are influential Nigerians from within and outside of the Niger Delta and include retired and serving high-ranking military officers and politicians (*Sun* 2009). Also included are foreigners such as Lebanese businessmen who operate across West Africa and are notorious for involvement in illegal businesses; Eastern Europeans (Poles, Russians, Ukrainians); and Filipinos (*Vanguard* 2009b). Crucially, the key destinations for stolen crude are mostly refineries in surrounding West African countries (Sierra Leone, Ghana, and Côte d'Ivoire) and international parallel markets.

An important development in the stolen crude market is the emergence in the last five years of a huge domestic, intra-Nigerian market, owing to persistent scarcity of refined petroleum products and increased use of power-generating sets. This has resulted in the mushrooming of local "refineries," which manually refine stolen crude to distill household-grade diesel fuel.

The scale of illegal bunkering

There is lack of clarity as to the exact scale of the problem. Available data appear low in reliability, consistency, and accuracy; however, they do suggest broad trends. The lack of credible data is due to either official attempts to cover up the actual scale of the problem or the use of different measurement yardsticks (such as a drop in estimated daily output or rough estimation of crude lost during production). The problem of unreliable data also stems from the fact that successive Nigerian governments lacked the capacity for accurate data on stolen crude oil.

According to media reports, the amount of stolen crude in 2000 was nearly 51,000 barrels; in 2001 it rose exponentially to over 262,000 barrels; in 2002 it dropped slightly to around 255,000 barrels (Coventry Cathedral 2009: 153). During 2003 and 2004, oil thieves stole an estimated daily average of between 250,000 and 400,000 barrels. In 2003, Shell Petroleum Development Company (SPDC) claimed to have lost an average of 100,000 barrels, and other upstream companies also reported losing an average of 150,000 barrels per day, all due to illegal oil bunkering (Coventry Cathedral 2009: 159–187).

In 2004, the president's special adviser on petroleum and energy, suggesting official estimates, stated that between 10 percent and 15 percent of Nigeria's daily output of 2 million barrels was being lost to illegal bunkering. At 2004 prices ($41), this was equivalent to US$8.5 million per day, and illegal bunkerers stole an annual average of US$4.2 billion (Coventry Cathedral 2009: 187). This puts the daily average of stolen crude at approximately 250,000 barrels, thus making 2003 the turning point in the expansion of illegal oil bunkering in Nigeria. Similarly, the International Maritime Organization estimated Nigerian cargo theft in 2006 to be 80,000 barrels per day, equivalent to US$1.6 billion for the year (Davis 2007: 3).

The illegal bunkering syndicate

The range of actors can be grouped into two broad categories: local and foreign. *Local actors* can be subdivided into five categories, depending on their roles in the chain of illegal oil bunkering.

1. *Armed groups and criminal gangs*: Despite the 2009 amnesty program, armed bunkering groups continue to exist in the Niger Delta. These groups derive huge revenue from the oil-bunkering racket through the collection of security clearance fees, passage fees, or protection fees. The participation of black market operators (in arms and stolen crude) from Eastern Europe is often a pointer to the connection between illegal oil bunkering and the proliferation of arms in the Niger Delta.
2. *Local communities*: Local communities or their agents have been known to be complicit in oil bunkering by either their failure to report the activities of oil bunkerers or their willing invitation to and derivation of revenue from oil bunkerers who operate in their communities. In most cases, crude pipelines that crisscross communities are sabotaged deep into community territories to give oil bunkerers the opportunity to hot-tap crude from pipelines. Some communities justify complicity in oil bunkering as a direct way of deriving financial benefit on oil extracted from their communities after years of neglect by government and oil companies.
3. *Colluding security personnel and other state officials*: Officials of state agencies, including the Nigerian National Petroleum Corporation and security Joint Task Force units, have been known to be deeply complicit in oil bunkering. There are reports of officials of Nigerian National Petroleum Corporation selling crude from that reserved for domestic consumption to black market operators and illegal vessels, and of their complicity in the falsification of bills of lading and theft of crude by licensed oil transporters. Security agencies continued to be plagued by bunkering-related corrupt practices through their taking of payments from bunkering syndicates. A 2003 *Report on Illegal Sale of Crude Oil* submitted to former President Ọbasanjọ is said to have detailed the explicit involvement of security officers (especially naval units based at NNS Okemiri) in the racket (Davis 2007: 2). The

involvement and dismissal of two rear admirals over the *MT African Pride* is a good illustration.

4. *Local artisans*: This group is composed mostly of youth that supply the needed manpower in the process of extraction, sourcing, storage, and transportation of stolen crude. This could range from welders, plumbers, and manual laborers to speedboat operators. For a majority of people in this category, oil bunkering provides employment and income, especially against the prevailing poverty in the Niger Delta.

5. *Local entrepreneurs*: This group is composed of local organizers ("businessmen") who coordinate the different phases and actors involved in oil bunkering. They make contact with armed groups, security agencies, and local communities; recruit local artisans; effect relevant payment; and find willing buyers for stolen crude.

Foreign actors can be subdivided into two categories, depending on their roles in the bunkering chain:

1. *Vessels crew/seafarers*: This group is diverse in nationalities, but the most notable are Filipinos, Russians, Ukrainians, and Ghanaians. They are usually crew members of barges and vessels used in transporting stolen crude. Up to ten Russians were arrested and later escaped in the 2003 *MT African Pride* episode. Also, in February 2009, 13 Filipinos pleaded guilty and were sentenced to various jail terms and fines for oil bunkering (*Nation* 2009a, b).

2. *International black market entrepreneurs*: This group operates at the topmost end of the chain, with or without any involvement in how and where stolen crude are sourced, but is interested only in having stolen crude delivered to their vessels and duly paid for. This group often includes Lebanese and Russian businessmen.

Policy responses

The Nigerian government has responded to the menace of oil bunkering through a mix of policy and operational capacity reforms.

1. *Enhanced security*: The government has taken two critical steps to enhance security in the Niger Delta region. First is the formation of a multiagency security Joint Task Force to stem the tide of insecurity, arrest militancy, and combat criminality (including sea piracy and oil bunkering) in the Niger Delta. It combines naval, army, air force, police, and intelligence components under a unified command structure. The government has also undertaken major

military acquisition, including aerial surveillance crafts, more naval patrol crafts, and other amphibious vehicles for the army to improve operational capabilities, as well as building new forward operating bases deep into the Niger Delta (Sahara Reporters June 25, 2008). Second is the negotiated settlement with the insurgency in the region, formalized under the October 2009 presidential amnesty program, and the disarmament, demobilization, and reintegration of militants. This has translated into improved security in the region in the short term; however, illegal bunkering has yet to abate. The two initiatives appear to be paying some dividends through increased detection and disruption of bunkering activities. Since January 2012, the Joint Task Force claimed to have recorded 1,653 arrests in 7,585 patrols, destroyed 3,778 homemade refineries and 120 barges, and seized several thousand pieces of bunkering equipment (*Punch* 2012).

2. *International proposal on blood oil*: In 2009, Nigeria made a major proposal at the 63rd UN General Assembly for an international action plan to curb the trade in stolen crude, noting that Nigeria's navy had seized approximately 260 ships over the previous three years, all containing stolen crude oil. The proposal calls for an oil certification ("fingerprinting") scheme, similar to that used for diamonds (Kimberley Process), which makes it difficult, if not impossible, to sell stolen crude. The proposal has yet to be reflected in any concrete policy initiative.

3. *Reducing incentives to trade in stolen crude in West Africa*: Since 2002, the Nigerian government, having identified neighboring countries in West Africa (Togo, Benin, Sierra Leone, Côte d'Ivoire) as attractive markets for stolen crude, engaged in bilateral agreements that involve the official sale of crude to them at discounted rates in exchange for repudiating the trade in stolen crude. This has been partially successful and has made the international (beyond West Africa) parallel market key destinations for stolen crude from Nigeria, thereby increasing operating costs and prices of bunkered oil. As more countries in West Africa become oil producers (due to recent oil discoveries in Ghana, Côte d'Ivoire, Sierra Leone, and Niger), it increases the prospects for a regional approach to addressing bunkering.

4. *Capacity building for government agencies*: Since 2009, the Nigerian government has taken steps to improve its capacity for monitoring and regulating the oil sector. This involves four key elements: first is the ascension to the Nigerian Extractive Industries Transparency Initiative (NEITI) to block revenue leakages and increase

accountability and transparency in the management of hydrocarbon resources in Nigeria. Second is the reform of the oil sector as proposed in the 2012 Petroleum Industry Bill, which is expected to transform the oil sector by addressing all forms of illegal bunkering through the creation of a Petroleum Host Community Development Fund and a new inspectorate division with special powers to undertake surveillance, investigate, and search and detain any persons, organizations, premises, and vessels suspected of involvement in illegal bunkering (FGN 2012: 41). The Petroleum Host Community Development Fund is to be funded by a contribution of 10 percent of net profits by oil multinationals operating in the upstream sector. Third is the September 2012 proposal to station Weights and Measures Departments in all oil export terminals. This is aimed at addressing official illegal bunkering by minimizing the risk of inaccurate disclosure of oil and gas exports (Punch 2012). Fourth is improving maritime security through the proposal to create a new Maritime Security Agency dedicated to addressing piracy and bunkering on the waterways (*Guardian* 2010; *Sun* 2010).

5. *Concessioning maritime operations to former militants*: One key, yet contentious, fallout of the 2009 amnesty program was the strategic concessioning of the country's waterways to Global West Vessel Specialist Agency (GWVSA), reportedly owned by a former militant leader (Government Tompolo) in 2012. The arrangement empowers GWVSA to provide platforms, security boats, equipment, and expertise to ensure maritime safety (and tackle illegal bunkering) and collect revenue on behalf of the government, in return for payment for its service once its revenue profile meets a set (undisclosed) threshold (*Vanguard* 2012a). A recent report that GWVSA led security agencies to bunkerers' hideout (Igbokoda) would appear to support the government's strategy of using insiders (as former militants all engaged in illegal bunkering) to checkmate the syndicate (*Vanguard* 2012b).

The policy responses since 2009 acknowledge the scale and seriousness of the problem and reflect a diversified approach to addressing the official and unofficial forms of bunkering. Thus far, some of the approaches appear to be working or have worked. These include the negotiated settlement of the armed conflict in the region through the amnesty program, which cut the blurred link between criminality and socioeconomic, political, and environmental grievances against the Nigerian state and oil multinationals. The amnesty program has led to improved security in the Niger Delta, reduced attacks on oil pipelines and infrastructures, and inevitably criminalized bunkering in all forms. Also noteworthy are improvements to the operational capabilities of the security agencies in the Niger Delta through increased surveillance and intelligence gathering. Since 2009, arrests, seizures, and prosecution for illegal bunkering have increased. The continued high level of stolen crude suggests that either the problem is bigger than first thought or it has increased on account of local and international demand (and prices) of crude. Finally, the bilateral initiative with neighboring countries in West Africa to create disincentives for bunkering could be a platform for an eventual regional approach to the problem.

However, the extant policy and operational responses—centrally the capacity-building effort—are ad hoc and lacking in proper planning, coordination, and building of synergies. For instance, more agencies are being created without proper definition and delineation of operational boundaries and strategies for interagency operability. Moreover, very little is achieved or achievable without a "whole of the government approach," for instance, by addressing the corrupt practices in the oil sector alongside activities among political elites. All extant responses are government led and lack the integration of civil society groups, their perspectives, and their independent oversight roles.

There is clearly a need for a regional and international approach to addressing illegal bunkering. While Nigeria has raised the issue at the UN, it has failed to follow up on it with concrete policy proposals and initiatives at the United Nations, African Union, and Economic Community of West African States (ECOWAS). A further complication is that some of the existing policy initiatives with huge potential to address illegal bunkering—in particular the NEITI—are voluntary in nature, with little legal obligation for action. For instance, successive NEITI audit reports indicating revenue leakages and failures of accounting system have attracted little or no government action. Finally, there is a disproportionate emphasis on a law enforcement approach to the problem, rather than alongside addressing the social, economic, political, and ecological undercurrents of the problem.

Comparative perspectives on criminality in the natural resource value chain in Africa

As reflected in the aforementioned case studies, there are five cross-cutting structural and proximate causes underlying criminal practices (theft, misappropriation, corruption, and side-stepping official processes) in the natural resource value chain.

First is weak capacity for official oversight of the natural resource sector. In Liberia and Nigeria, there is a genuine lack of capacity by the relevant government agencies to effectively monitor, regulate, document, investigate, detect, and impose sanctions where and when criminal practices occur in the natural resource value chain. In some cases, the basis of this lies in outdated and/or inadequate laws and regulatory regimes, with criminal syndicates either adopting more sophisticated methods to evade official detection or resorting to co-opting relevant officials.

Second is that where and when some measure of capacity exists, it is easily compromised by the active collusion of state officials. This underscores official corruption as a critical factor in the perpetuation of criminal acts in the natural resource value chain.

Third is poverty and material deprivation of the majority of local communities where natural resources are extracted. More often than not, the poverty levels in such communities are much higher than the national average, which makes resorting to a "self-help" strategy attractive to individuals in those communities.

Fourth is insecurity and violence signposted by the existence and activities of armed groups. The onset, intensification, and sophistication in the criminal exploitation of natural resources have paralleled the proliferation of small arms and light weapons. Armed conflicts often mask the illegal tapping of natural resources and blur the divide between criminality and genuine protest against socioeconomic, political, and environmental injustices.

Finally, globalization of economic production, marked by the increased demand for Africa's natural resources—generally referred to as the "new scramble for Africa" (Obi 2009)—and the resulting increase in prices have emerged as key variables in the expansion in criminal practices in Africa's natural resource value chain. From a yearly average of approximately US$28 in 2000, Nigeria's low-sulfur Bonny Light crude increased to a monthly average of nearly US$75 in 2007 and US$115 in 2008; it has hovered near the US$100 mark ever since. The increase in criminal practices in Africa's natural resource sector appears to have started from the 1990s, a period synonymous with increased globalization.

In terms of actors, the criminal syndicates profiteering from the natural resource value chain in Africa tend to involve at least four critical elements. First are colluding government officials and influential political elites who provide official cover and underwrite the institutional, political, and bureaucratic costs of criminal practices in the natural resource value chain. Second are colluding state security agencies and armed groups who profiteer either directly (through active exploitation themselves or by way of their appointed agents) or indirectly (through collection of illegal taxes, rents, and commissions) from criminal practices. Either way, this category underwrites the security costs of criminal practices in the natural resource value chain. Third are international criminal networks that steal or buy stolen natural resources for onward marketing on the international parallel market at reduced prices. This group covers nationals from African and non-African countries. Fourth are local communities and their inhabitants who encourage, condone, or participate in illegal exploitation of natural resources. In a majority of cases, local communities' involvement is self-help strategy: a reaction to official neglect, poverty, material deprivation, corruption, and environmental damage arising from official extraction of natural resources.

The impacts of criminal practices in Africa's natural resource value chain are at least fourfold. First is the apparent loss of official revenue as criminal networks and practices undercut both the capacity to collect, and the level of, revenue accruing to the government. Second is the onset or transformation of grievances into violence or the escalation of existing armed conflicts. Levels of violence tend to rise with increases in criminal practices in the natural resource value chain in Africa; criminality in the natural resource value chain often mutates into armed conflicts and vice versa. Third are the cross-border and regional dimensions and impacts. In Liberia and Nigeria, criminal practices in the natural resource value chain encompass cross-border collaboration and region-wide networks to transport and sell stolen natural resources. In addition, the insecurities generated by profiteering from the natural resource value chain do spill across borders, as underscored by the cross-border contagion effect of the conflict in Liberia in the Mano basin area and by the spread of piracy in West African waterways. Fourth are the often-hidden long-term impacts on the local communities, marked by the distortion and disruption of local economies (agriculture) and damage to environmental sustainability.

The landscape of policy responses to criminal practices in the natural resource value chain in Africa is characterized by at least five observations. First is that the core of policy responses have a national base, notwithstanding emerging international civil society-led initiatives such as the extractive industries transparency initiatives and the Publish

What You Pay (PWYP) schemes. There is limited synergy between national and international policy responses to criminality in the management of most natural resources in Africa. Second, current responses continue to be marked by the lack of genuine capacity to properly translate existing laws and policies into action or match the increasing sophistication of criminal syndicates profiteering from Africa's natural resource value chain. Complex bureaucracies, lack of clarity, duplication of agencies and functions, limited coordination, and corruption also beset a majority of extant policy responses. Third is the focus on formal, government-led responses. Despite the imperatives of formal processes, there remain crucial limitations related to the need for formal processes to respond to or capture unofficial practices. In short, there are gaps between legal, official approaches and informal practices connected with sociocultural practices and unresolved contestations over the ownership and control of natural resources in a majority of cases. Fourth is the lack of a robust approach to addressing profiteering from the natural resource value chain by transcending law enforcement and addressing underlying socioeconomic, political, environmental, and cultural issues. Fifth is the absence of concrete regional approaches and policies (notwithstanding fleeting mentions and debates at the ECOWAS).

In the light of the foregoing, it is possible to highlight elements of extant policy responses that have been proven effective or have the potential to address criminality in the natural resource value chain:

1. A whole range of government approaches that integrate the management of criminality in the natural resource value chain into overall development and security planning. The postwar GEMAP initiative reflects a semblance of this approach, as it coordinates national and international initiatives and represents a multi-sectoral development planning scheme in the management of economic resources, including natural resources.
2. The scale and visibility of criminality in the natural resource value chain greatly reduces as security improves. Accordingly, enhanced security represents a critical element of successful efforts to address criminality in the natural resource value chain. The negotiated resolution of armed conflicts in the Niger Delta region and in Liberia provided the platform to seek improved regulation of the natural resource value chain.
3. The imperative of local perspectives and participation of civil society groups are crucial to the success of sustainable efforts at addressing

profiteering within the natural resource value chain in Africa. There is a distinction between a government-led approach and local ownership; policy initiatives must reflect and be anchored in the perspectives, concerns, and interests of local communities and civil society groups. More important is the need to recognize informal sociocultural practices and approaches to natural resource management in host communities and empower locals through incentives to move them unto official realms (rather than simply outlawing and criminalizing their practices).

4. It is essential to develop and implement a regional approach that outlines normative and practical policy frameworks for addressing criminal practices and overall natural resource governance in Africa. The prospects and gains from this are numerous and include opportunities for shared learning, coordination of policies and practices, and regional law enforcement (anti-criminal) systems in the natural resource value chain. Existing bilateral agreements, regional integration schemes, and the African Union's Common African Defence and Security Policy (CADSP) framework provide platforms for a regional approach and for transforming normative frameworks into practical policy initiatives. Similarly, existing voluntary initiatives (Extractive Industries Transparency Initiative and PWYP), already used at national levels, could be adapted and adopted as regional natural resource governance systems.

Conclusion

This chapter explored the complex network of issues, processes, and actors involved in criminal practices in the natural resource value chain in Africa. It identified underlying causes to include weak oversight mechanisms, genuine capacity deficits, corruption and collusion by government agencies and officials, gaps between legal (official) and informal (sociocultural) approaches, unresolved contestation about natural resource governance, armed conflicts and insecurity, and globalization. The criminal syndicates involved were highlighted to include colluding local communities (and their inhabitants), government officials, security agencies, and foreign parallel market operators. The chapter also noted the cross-border and regional dimensions and implications of criminality in the natural resource value chain in West Africa. The listed impacts include loss of revenue to governments, mutation into armed conflict (and vice versa), contagion effect across borders and regions, and long-term damages to local socioeconomic and environmental sustainability. Most importantly, the chapter noted the limitations of extant policy responses to include narrow focus on law enforcement, rather than alongside addressing underlying socioeconomic, political, and environmental issues; strong emphasis on national-level initiatives with limited synergies with international approaches; and the absence of strong regional and international policy actions. Finally, the chapter pinpointed extant resources and initiatives either working or with potentials to contribute to addressing profiteering from the natural resource value chain in Africa. This includes a whole-of-the-government approach, enhanced national and regional security through negotiated settlement of disputes, and other conflict prevention initiatives; region-level adaptation and adoption of emerging best practices in the management of natural resources; and the integration of local perspectives and practices (rather than simply criminalizing them) and CSOs into official policy responses.

Notes

1. UNSC (2003) Resolution 1478 S/RES/1478 (2003), May 6. Other relevant UNSC resolutions on Liberia include: UNSCRs 1306 (2000); 1689 (2006), and 1819 (2008).
2. See US Energy Information Administration (2012).
3. Official investigations and convictions (payment of fines) of western companies involved in Nigeria's oil and gas sector, including Siemens and Halliburton, by their home governments since 2008 have revealed huge levels of corrupt practices, especially the bribing of high-level state officials, including the presidency and the ruling PDP in return for the award of contracts. See Vanguard (2010).

References

Alao, C. (2007). *Natural Resources and Conflict in Africa: The Tragedy of Endowment*, New York: University of Rochester Press.

All Africa News. (2009). "Ghana Seize Nigerian Oil Vessel," All Africa News, November 28, Available at: http://allafrica.com/stories/200911300451.html. Accessed September 13, 2012.

Beevers, M. (2012). "Forest Resources and Peacebuilding: Preliminary Lessons from Liberia and Sierra Leone," in P. Lujala and S.A. Rustad (eds.), *High-Value Natural Resources and Peacebuilding*, London: Earthscan.

Busch, G. (2005). "Whom Do They Think They Are Fooling?" Available at www.dawodu.com/busch1.htm. Accessed August 3, 2012.

Cohen, N., Mohan, C., Woiwo, K., Whawhen, J., Dahoh, S., and Snelbecker, D. (2010). "Final Evaluation of USAID GEMAP Activities," *USAID Report Prepared by Sibley International LLC*.

Coventry Cathedral. (2009). "The Potential for Peace and Reconciliation in the Niger Delta," Available at: http://www.coventrycathedral.org.uk/downloads/publications/35.pdf. Accessed August 10, 2012.

Daily Sun (2009). "After Militancy, JTF Tackles Oil Bunkering," November 3, Available at: http://sunnewsonline.com/new/. Accessed September 4, 2012.

Davis, G. (2007). "Shifting Trends in Oil Theft in the Niger Delta," *LegalOil.com Information Paper* 3, January, Available at: http://www.legaloil.com/Documents/Library/Legal%20Oil%20Information%20. Accessed September 1, 2012.

Ellis, S. (1999). *The Mask of Anarchy: The Destruction of Liberia and the Religious Dimension of an African Civil War*, New York: New York University Press.

Federal Government of Nigeria (FGN). (2012). "Draft Petroleum Industry Bill, 2012," Available at: http://www.proshareng.com/reports/5071/The-New-PIB-Draft-Bill-2012. Accessed September 16, 2012.

Forest Law Enforcement Governance and Trade. (2012). "Civil Society Organisations Join EU Battle against Illegal Logging," *Briefing Note*, July.

Global Witness. (2001). "Liberia Breaches UN Sanctions whilst Its Logging Industry Funds Arms Import," Available at: http://www.globalwitness.org/library/liberia-breaches-un-sanctions-whilst-its-logging-industry-funds-arms-imports-and-ruf-rebels. Accessed August 10, 2012.

Global Witness. (2012a). *Spoiled: Liberia's Private Use Permits*, August, Available at: http://loggingoff.info/sites/loggingoff.info/files/Spoiled%20-%20Liber. Accessed September 1, 2012.

Global Witness. (2012b). *Signing Their Lives Away: Liberia's Private Use Permits and the Destruction of Community-Owned Rainforest*, September, Available at: http://www.globalwitness.org/sites/default/files/library/Signing%20the. Accessed September 27, 2012.

Globserver. (2012). "Nigeria Loses $1bn a Month to Oil Theft," September 17, Available at: http://www.globserver.cn/en/nigeria/press/nigeria-loses-1bn-month-oil-theft-2012-09-17. Accessed September 18, 2012.

Guardian. (2010). "National Assembly Moves to Clear Cloud over New Maritime Agency," January 13, Available at: http://www.ngrguardiannews.com/. Accessed September 18, 2012.

International Crisis Group. (2003). "Tackling Liberia: The Eye of the Regional Storm," *Africa Report No. 62*, April 30.

Nation. (2009a). "13 Filipinos Bag 65 Years Imprisonment for Oil Theft," February 23, Available at: http://thenationonlineng.net/new/. Accessed September 2, 2012.

Nation. (2009b). "Two Ships for Navy," March 12, Available at: http://thenationonlineng.net/new/. Accessed September 10, 2012.

Obi, C. (2009). "Scrambling for Oil in West Africa," in R. Southall and H. Melber (eds.), *The New Scramble for Africa: Imperialism, Investment and Development in Africa*, KwaZulu-Natal: University of KwaZulu-Natal Press.

Punch. (2012). "FG's Plan for Crude Exports Creates Disquiet in Oil Sector," *Punch Online*, September 30, Available at: http://www.punchng.com/news/fgs-new-plan-for-crude-export-creates-disquiet-in-oil-sector/. Accessed September 30, 2012.

Reno, W. (2000). "Shadow States and the Political Economy of Civil Wars," in M. Bergdal and D.M. Malone (eds.), *Greed and Grievance: Economic Agendas in Civil Wars*, Boulder, CO: Lynne Rienner.

Richards, P. (2001). "Are Forest Wars in Africa Resource Conflicts? The Case of Sierra Leone," in N.L. Peluso and M. Watts (eds.), *Violent Environments*, New York: Cornell University Press.

Sahara Reporters. (2008). "Niger Delta: Obasanjo Acquires 193 Cobra Amphibious Armored Vehicles," June 25, Available at: http://www.saharareporters .com/index.php?option= com_content&view=article&id=1213:niger-delta-obasanjo-acquires-193-cobra-amphibious-armored-vehicles-&catid=1:latest-news&Itemid=18. Accessed August 17, 2012.

Sun. (2010). "NSA, NAVY, Akhigbe Disagree over Maritime Safety Bill," January 14, Available at: http://sunnewsonline.com/new/. Accessed September 16, 2012.

ThisDay. (2008). "Customs Can't Give Accurate Figures on Oil Lifting," August 9, Available at: http://www.thisdaylive.com/. Accessed September 10, 2012.

TRAFFIC. (2012). "Timber Trade," Available at: http://www.traffic.org/timber-trade/. Accessed December 24, 2012.

UNODC. (2010). *The Globalisation of Crime: A Transnational Organised Crime Threat Assessment*, New York: United Nations Publications.

UNSC. (2006). UNSC Resolution 1689 adopted on 20 June 2006, Available at: http://www.worldlii.org/int/other/UNSCRsn/2006/38.pdf. Accessed September 17, 2012.

US Energy Information Administration. (2012). *Country Brief Analysis: Nigeria*, Available at: http://www.eia.doe.gov/emeu/cabs/Nigeria/Oil.html. Accessed September 17, 2012. Available at: http://allafrica.com/stories/200903310259 .html. Accessed September 30, 2012.

Vanguard. (2009b). "The Foreign Connection in Oil Theft," January 10, Available at: http://allafrica.com/stories/200901120453.html. Accessed September 18, 2012.

Vanguard. (2010). "Halliburton: FG Withdraws Charges against Jeffrey Tesler," October 14, Available at: http://www.vanguardngr.com/2010/10/halliburton-fg-withdraws-charge-against-jeffrey-tesler/. Accessed October 15, 2010.

Vanguard. (2012a). "Why We Awarded Waterways Contract to Tompolo— FG," March 20, Available at: http://www.vanguardngr.com/2012/03/why-we-awarded-waterways-contract-to-tompolo-fg/. Accessed March 21, 2012.

Vanguard. (2012b). "Jitters in the Oil Sector amid Crackdown on Thieves, Pirates," September 30, Available at: http://www.vanguardngr.com/2012/09/jitters-in-the-oil-sector-amid-crackdown-on-thieves-pirates/. Accessed October 1, 2012.

5

Structured Transformation and Natural Resources Management in Africa

William G. Moseley

Introduction

Despite a dip during the global economic recession, many African nations continue to ride a natural resource and commodity export-driven economic boom. A critical question for these countries is whether this boom is somehow different than those of the past, which all faded with time. Some have argued that we are experiencing an underlying shift in global demand for commodities because of growing urbanization on the international scale and the swelling ranks of the middle class in the global South. These factors are seen as having fundamentally changed the demand for food and raw materials that many African nations are well placed to supply. Such deeper shifts in demand may lead one to conclude that African economies, which are largely dominated by primary production and are the least diversified of all developing regions, should merely ride the wave to sustained prosperity. Furthermore, some have posited that African economies ought to remain focused on primary production, the area in which they have an advantage relative to other actors in the global economic system (Naude et al. 2010).

This line of thinking neglects the longer-term trend of declining real commodity prices. This is because shorter-term rises in prices almost always lead to the introduction of new sources of production, substitutes, or improved efficiency of use. Worse yet are long-range declining terms of trade wherein the price fetched for African exports (dominated by primary production) has weakened relative to the costs of goods imported. This evidence leads to a different conclusion than that noted in the previous paragraph: African economies must diversify away from primary production (resource extraction and agricultural commodity production) to improve their economic position vis-à-vis the rest of the world. Economies overly focused on primary production are risky because they tend to lack diversity and are vulnerable to fluctuations in global commodity prices. Such economies also often miss out on value-added processing of raw materials and other forms of manufacturing.

Do natural resources intrinsically impede economic diversification? Under what conditions can resource-rich economies diversify, and how can these conditions be created in Africa? Finally, what lessons and experiences can be adapted for use by countries seeking to use natural resource wealth to fuel development?

Understanding trends in African resource-based economies: Recent trends in mineral, energy, and agricultural exports from African countries

Many African economies sustained above average levels of growth through most of the international economic recession that began in 2008 as they rode a cyclical commodity export boom.[1] Leading up to 2008, most of Africa was in the midst of one of its highest growth periods on record, with average (real) gross domestic product (GDP) rising by 4.9 percent a year from 2000 through 2008, more than twice the rate in the 1980s and 1990s (Leke et al. 2010). Real GDP dipped to 3.1 percent in 2009, at the height of the financial crisis, but rose again to 4.9 percent in 2010. While average growth slackened to 3.7 percent in 2011 as a result of sociopolitical unrest in some African countries, it is expected to recover to 5.8 percent in 2012 (African Development Bank Group 2012). In comparison, global (real) GDP averaged 4.3 percent per year during 2000–2008. It then dropped to −0.7 percent in 2009 and rose to 4.9 percent in 2010 and 3.7 percent in 2011 (CIA 2012). As such, African economic expansion over the past 12 years has outpaced that of the world over the same period, as well as its own growth in the previous two decades.

Rising prices for oil, minerals, and other commodities have helped lift African GDPs since 2000 (Leke et al. 2010). For starters, the continent is well endowed with certain natural resources, including oil (10 percent of global reserves), gold (40 percent), chromium, and platinum (90 percent) (Leke et al. 2010). Furthermore, much is made of the fact that large areas of Africa have not been subjected to prospecting for mineral resources, so there is a belief that much larger reserves may exist (Elliot et al. 2011).

In terms of land, and related production of food and fiber, it is debatable whether Africa really has excess food and fiber to export, given that it struggles to feed its own population in some years. That said, there is a perception by some outside powers that Africa has plenty of land to spare (Horta 2009). In fact, this perception of "excess land" has been used to legitimize long-term land leases or "land grabs." The 2012 *Africa Capacity Indicators* report on agricultural transformation and food security discussed this problem at length and concluded that foreign acquisition of African land affects the land rights of the poor and women and that it has implications for capacity building (ACBF 2012).

One estimate is that roughly one-third of African economic growth is tied directly to commodity exports (including oil, minerals, and agricultural goods) (Roxburgh et al. 2010). The dominance of commodities in African economies is even more evident in export figures, as approximately 65–80 percent of the continent's exports are composed of such goods. In fact, commodities exceed 50 percent of exports for nearly every African country; in some cases (such as Algeria, Nigeria, Angola, and Kenya), they account for over 90 percent of a country's export revenue (Oramah 2012). Commodities constitute a larger share of exports in Africa than in any other developing region, and the proportion has been growing over time (Figure 5.1). This growing dependence on commodity exports stands in stark contrast to other areas of the world that have developed rapidly while diversifying their economies (Gelb 2010). Given the dominance of commodities in Africa's export and growth boom of the 2000s, some have framed this situation as "the new scramble for Africa," with an apparent reference to the first scramble for Africa in the late 19th century (Carmody 2011). While growth began to slow for some sectors as the global downturn persisted, it has now begun to turn around, and projections are for consumption of many commodities to increase by 25 percent over the coming decade, approximately at twice the rate of the 1990s (Leke et al. 2010).

Underlying global demographic, economic, and market shifts that shed light on export trends

The current commodity-driven export boom is not the first experienced by African countries; similar booms occurred in the 1950s and 1970s. The real question is whether this boom is different—which inevitably pushes us to ask what accounts for the rise in global commodity

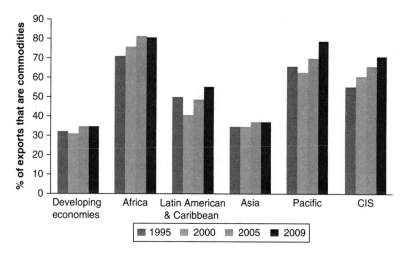

Figure 5.1   Share of commodities in total exports by developing region
Source: United Development Programme (UNDP). 2011. "Chapter 2: Commodity Dependence and International Commodity Prices." In: *Towards Human Resilience: Sustaining MGD Progress in an Age of Economic Uncertainty.* New York: UNDP. Permission granted.

consumption and whether this is an episodic or a more basic shift. Furthermore, we are prompted to consider what Africa can do differently this time around to ensure that the benefits from the boom lead to development and poverty reduction.

Some have asserted that the current boom is fundamentally different than those of previous periods. The president of the African Development Bank Group, Donald Kaberuka, has argued (2007) that this commodity boom is different because, unlike the in 1970s, oil demand in Africa today is not being driven by cartelization (such as the Organization of Petroleum Exporting Countries) or a sudden geopolitical shock (like the Iran-Iraq War), but rather by more fundamental shifts in global demographics and demand. Furthermore, growth in the current period is not linked to industrial expansion in the global North (such as that which occurred after World War II) but rather to phenomenal growth in other areas of the global South.

Demand for commodities is growing fastest in the world's emerging economies, particularly in Asia and the Middle East. This demand is driven in part by urbanization and related changes in consumption patterns in these countries (Leppman 2005) (see also Figure 5.2). To a lesser extent, urbanization within Africa is also driving demand for commodities. In the case of Asia, and most particularly China, not only has the

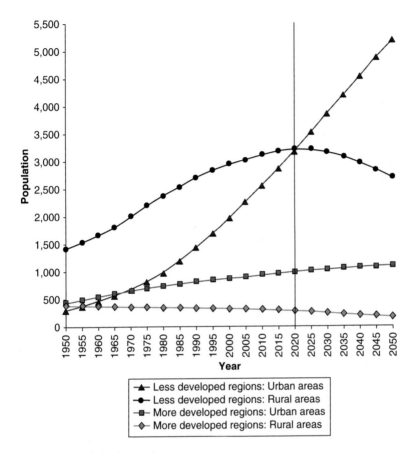

Figure 5.2 Global urbanization patterns
Source: United Nations. 2010. *World Urbanization Prospects: The 2009 Revision*. ESA/P/WP/215. New York: UN Department of Economic and Social Affairs, Population Division. Permission granted.

country's populace become more urban, but the economy is increasingly dominated by manufacturing. China's economy has effectively moved from the periphery to the semi-periphery and, as the world's great factory, it must source raw materials from around the world (Muldavin 2007; Bergmann 2012). As a result of these changing global demographics and industrialization patterns, Africa's trading patterns have shifted geographically.

Despite its history of commercial ties with Europe, Africa now conducts half its trade with other developing economic regions of the world.

For example, from 1990 to 2008, Asia's share of African trade doubled to 28 percent, whereas Europe's portion shrank from 51 to 28 percent (Leke et al. 2010). Ironically, Africa has not always been able to meet the commodity demands of its own growing urban population. For example, growing urbanization in West Africa has led to a surging demand for rice, much of which is provided by Asian producers. A combination of poor internal infrastructure, cheap broken rice from Vietnam and Thailand, few to no tariff barriers, and limited state support for agriculture has made it difficult for local producers to compete (Moseley et al. 2010).

A contrary view is that this is just the upside of a cyclical commodity boom-and-bust cycle, and that nothing fundamental has changed. While Africa has benefited from the current commodity boom, such high returns will soon decline. In fact, recent reports suggest that we are nearing the end of a commodity super-cycle (Sharma 2012). Evidence for this includes indicators of "oversupply, falling prices and mine closures in markets as diverse as oil, diamonds, platinum and ferrochrome" (GGA 2012a: 25).

Analysts point to long-term trends of commodity prices generally declining over time except for oil (Figure 5 3). When prices go up, other suppliers emerge, substitutes are developed, and new levels of efficiency

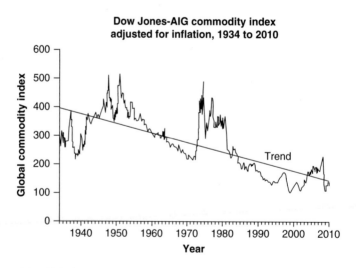

Figure 5.3 Global commodity index adjusted for inflation
Source: Mark Perry, Carpe Diem blog (http://mjperry.blogspot.com/).

are attained. In some instances, subsidies may also be encouraging overproduction in the global North.

Agricultural commodities are clearly problematic, as large numbers of countries around the world are capable of producing them. In the case of tropical crops, other developing countries with relatively inexpensive labor costs may competitively enter markets as they believe there are profits to be made, or they may perceive that commodity production is one of their only economic development options. This was arguably the case for Vietnam, which saw its robusta coffee production and exports surge from the late 1980s, creating havoc for Africa's major robusta coffee producers, Uganda and Côte d'Ivoire (USDA 2012).

Other types of commodities are produced in both temperate and tropical zones. Cotton is a good example of an African cash crop in this category. These crops may be produced not only in other tropical countries where labor is cheap, but also in countries of the global North where production may be subsidized for political reasons. Oxfam (2003) determined that in the 1990s, at the height of American cotton subsidies, the six African countries that depended on cotton for more than 20 percent of their total revenue (Benin, Burkina Faso, Central African Republic, Chad, Mali, and Togo) increased export volumes by 40 percent, yet saw their export revenue decline by 4 percent. African cotton producers and their allies in other regions of the global South did successfully win a suit they lodged against the United States with the World Trade Organization (WTO). However, while the United States did meet the letter of the law to settle this complaint, it simply switched the form of subsidy payments it provided to its farmers with no real consequent impact on global cotton prices (Ledermann and Moseley 2008). Furthermore, Africa is at an even further disadvantage as a raw cotton exporter attempting to compete with producers located in closer proximity to production facilities because it lacks the capacity for first- and second-stage transformation of products such as cotton (although it did have that capacity in some cases before the structural adjustment period) (Carmody and Taylor 2003). The irony, of course, is that Africa imports most of its textiles, goods for which it often produces the raw material.

We tend to think of mineral and energy resources as nonrenewable and finite, which suggests that, at least in this category of commodities, the number of suppliers and degree of competition will be limited. Yet, even in these cases, rising prices tend to be temporary as technological efficiencies are attained, new reserves brought into production (often made available by technological breakthroughs), and substitutes developed. A good example of this is natural gas, for which

Algeria, Nigeria, Egypt, and Libya are major African exporters. Now the competition is increasing on the continent with new discoveries in Mozambique, Tanzania, and Uganda. Even in well-prospected areas, new technologies are opening up vast new supplies For example, due to the environmentally controversial process known as hydraulic fracturing (popularly known as fracking), and especially a newer technique known as horizontal slick-water fracking, huge new reserves of natural gas have been opened up in the United States and Canada and have depressed global prices. When adjusted for inflation, natural gas prices are now as low as they were in the mid-1970s (Tverberg 2012).

On top of declining real prices (over the longer term) for most of the world's major commodities, there have been noticeable and consistent declines in terms of trade for African commodity crops over time. In relative terms, the money Africa is earning from its commodity exports is enabling it to import less from other parts of the world.

Foreign involvement in Africa's resource-based economies

In terms of exports of African commodities, the United States and China are by far the most important destinations, followed by France, the United Kingdom, India, and Brazil (Carmody 2011). To some extent, there is competition between the United States and China as they scramble to secure access to energy resources (Carmody and Owusu 2007). This competition can be said to have allowed some African states to negotiate better deals. It has also led to aid packages and infrastructure projects that are directly tied to resource access grants. This new form of aid, in which the private sector and bilateral donors work in tandem, was introduced and honed largely by the Chinese.

China, for example, bid for access to 10 million tons of copper and 2 million tons of cobalt in the Democratic Republic of the Congo in exchange for a US$6 billion package of infrastructure investments, including mine improvements, roads, rail, hospitals, and schools. These types of arrangements are now also being negotiated between African governments and those of India, Brazil, and several Middle Eastern states (Leke et al. 2010).

Risks of economies overly focused on primary production: The external and internal vulnerability of undiversified economies

Global commodity prices are notoriously volatile. This need not be a problem for a national economy if it is sufficiently diversified, across both different types of commodities in the primary sector and different types of economic activity—primary, secondary, and tertiary production. However, when an economy is overly dependent on the export of one or two commodities, it becomes hostage to such vulnerability, not to mention long-term downward trends in real prices for most of the world's major commodities.

Scholars like Collier (2007) have also argued that there is a difference between undiversified economies based on the extraction of oil, gas, and minerals and those based on the production of agricultural products. For oil, gas, and minerals, wealth generally accrues to a small group of specialized workers, foreign investors, and the state. The state's share of these returns, as well as its ability or desire to redistribute or reinvest this revenue, also varies greatly. In contrast, for agricultural products, the returns tend to be distributed to a much broader workforce, as well as foreign investors and the state. "Nearly all of Africa's approximately six million commercial farm workers, and a very large proportion of its 140 million smallhold farmers, have some involvement with cash crop production" (Moseley and Gray 2008: 3). But the distribution of revenue from cash crops also tends to vary based on the major mode of production for different commodities (as some are grown primarily by smallholders, while others are produced mainly on large plantations). Given the quicker return on investment, annual crops such as cotton and tobacco tend to be more common on small farms, while tree crops, such as rubber and oil palm, are more likely to be grown on plantations. Tea, cocoa, and coffee appear on both plantations and smallhold farms. State and private sector involvement with these different crops also tends to vary greatly across the continent and thus has implications for revenue capture and political interest in the fortunes of different sectors. For example, the state has historically been heavily involved in the cotton sector in a group of former French colonies in West Africa, while private companies have largely taken the lead in East and Southern Africa.

Leke et al. (2010) have divided African economies into four groups based on the level of diversification within their economies. The wealthiest of these groups is composed of Africa's "wealthy oil and gas exporters," which have the continent's highest GDPs per capita. This group includes Algeria, Angola, Chad, Congo, Equatorial Guinea, Gabon, Libya, and Nigeria. Among this group are those that have been exporting oil for many decades and those that are relatively new to the business. Interestingly, this group of economies is also the least diversified among the four groups. Despite the narrow nature of their economies, rising energy prices over the past decade have likely muted any sense of urgency for diversification. For example, rising oil prices meant that the three largest producers (Algeria, Angola, and Nigeria) earned US$1 trillion from petroleum exports from 2000 through 2008, compared with US$300 billion during the decade of the 1990s.

In contrast to the first group, the second group of African economies is nearly as wealthy, yet its economies are much more diversified. The four "wealthier and diversified" countries in this group, Egypt, Morocco, South Africa, and Tunisia, have diversified their economies away from primary production and into secondary (manufacturing) and tertiary (services) activities, which account for 83 percent of their combined GDP. "Domestic services, such as construction, banking, telecom, and retailing, have accounted for more than 70 percent of their growth since 2000" (Leke et al. 2010).

In addition, there are two transitional groups. The first consists of Cameroon, Ghana, Kenya, Mozambique, Senegal, Tanzania, Uganda, and Zambia.[2] Overall, this group is not as wealthy as the first two groups discussed (as measured by per capita GDP), but among the two transitional groups it is diversifying at a faster rate, and we refer to them as "transitional but highly diversified." That said, within this group there are also varying levels of economic diversification. For example, some countries, like Zambia and Mozambique, are still heavily dependent on one commodity (copper in the former and aluminum in the latter), whereas others, like Kenya, Ghana, and Uganda, are already more diversified.

The second transitional group, and the last of the four clusters identified by Leke et al. (2010), is still relatively poor and includes countries like the Democratic Republic of Congo, Ethiopia, Mali, and Sierra Leone. Some of these countries, such as Ethiopia and Mali, are relatively poor in terms of natural resources and have large rural populations, and we refer to them as "transitional and poor in natural resources." Others are in the process of recovering from wars, such as the Democratic Republic of Congo and Sierra Leone, but have much wealthier resource endowments. All of these countries have grown steadily in the 2000s, but their growth is precarious and erratic.

Case studies of export economies focused on a few agricultural, mineral, or energy exports

This section explores three countries whose economies are heavily dependent on the production of one or two commodities for export. In selecting these countries, we looked for a range of income levels (from poor to wealthy by African standards) as well as a range of commodities (agricultural, mineral, and energy). These country studies are Mali (with a per capita GDP of $599) and its focus on cotton, Botswana (with a per capita GDP of $7,403) and its focus on diamonds, and Angola (with a per capita GDP of $4,322) and its focus on oil (UN 2010).

a) Mali and cotton

Prior to the recent political troubles in Mali (Moseley 2012), the World Bank had long encouraged the government to further develop its exports (mainly cotton) as part of its structural adjustment program. While Mali has other exports, such as trade in gold and cattle on the hoof to neighboring countries, cotton has long accounted for the largest share of goods leaving the country. With an annual average of 200,000 tons of production during 1998–2005, Mali was the largest cotton exporter in Africa for several years (FAO 2006). With cotton accounting for the largest share of GDP and government revenue, the Malian state was interested in maintaining and increasing cotton-related exports and revenue. Cotton held such a prominent position in the Malian political economy that it was sometimes referred to as *l'or blanc* (white gold) (Tefft 2004). Others noted that *"coton est le moteur du développement"* (cotton is the motor of development; GoM 1998), and became a sort of cure-all for the country's woes. Government officials argued not only that cotton promotes economic growth, but also that it enhances food security and promotes environmental stewardship. When interacting with farmers, for example, agricultural extension agents often asserted that *"kori tigi ye nyo tigi ye,"* which means that successful cotton farmers are successful millet farmers (Moseley 1993). In a variety of interviews with mid- and high-level government and donor officials in 2000, these persons suggested that poverty is a major source of environmental degradation. Cotton, they argued, provided the wealth necessary for farmers to be good stewards of the land.

Unfortunately, a long-term development strategy for Mali based on continued increases in cotton production was fraught with problems for the national economy as well as local ecologies and livelihoods. As global cotton prices declined, so did the country's export revenue. In many years, it was often irrational for small farmers to grow cotton, but they did so to have access to credit. As the state or donors were not willing to actively support agricultural diversification, the over-farming of cotton became deleterious to soil quality, and yields began to decline in the oldest production areas (Moseley and Gray 2008).

Mali was also unable to diversify its economy along the cotton value chain, although, to be fair, there are historical examples of such a value chain. For example, there was a lively, traditional weaving sector that was destroyed during the French colonial period (Roberts 1996). Furthermore, a local textile industry was built up during the import substitution era of the 1960s and 1970s. Although the state-run textile industry known as COMATEX was never really that competitive, its demise was cemented by a combination of structural adjustment, the international secondhand clothing trade, and cheap Chinese imports (Baden and Barber 2005). In contrast, the similarly endowed Burkina Faso has been more successful at developing a value chain from its cotton farmers to the manufacture of cotton pajamas and other organic cotton products (Bassett 2010).

b) Botswana and diamonds

Botswana is considered to be an African success story and has been heralded as an African development miracle (Samatar 1999). At independence in 1966, Botswana was suffering from a drought and had less than five kilometers of road, little to no physical infrastructure, and 23 college graduates. In fact, the country did not have a capital city, as the protectorate had been administered from Mafeking in South Africa. No one could have predicted the economic growth that was to come. The discovery of diamonds a few years after independence, and the astute management of this resource, allowed the country to invest heavily in education, infrastructure, and health care. These investments, combined with a relatively small national population of 2 million people, allowed the country to become one of Africa's most well to do. Botswana has been one of the fastest-growing countries in the world over the past 40 years in economic terms. This rate of growth is comparable to that of the newly industrialized countries (NICs) of East Asia, yet Botswana did this under democratic governance. Today Botswana has a per capita GDP of $13,100 and US$8 billion in reserves, enough to cover imports for 20 months. It is the world's leading exporter of precious diamonds.

However, despite all of its success, Botswana's economy is still narrowly dependent on diamonds, as these precious gems account for 76 percent of export revenue, 45 percent of government revenue, and 33 percent of GDP. Hillbom (2008) among others has argued that while Botswana may have experienced tremendous economic growth, it has not developed. For Hillbom, Botswana's "pre-modern growth, as opposed to development, allows for significant poverty rates and extremely unequal resource and income distribution to prevail in the midst of plenty" (Hillbom 2008: 191).

Unfortunately, mining employs only 4 percent of the labor force and is not complemented by other forms of industry. It mainly has spin-off effects on the public sector via government employment and spending. An unfortunate result of an economy narrowly focused on diamond mining that produces few jobs, and a government sector that largely employs highly educated urbanites, is that Botswana is a deeply unequal country. In fact, with a Gini index of 60, Botswana has one of the most unequal income distributions in the world. This inequality means that although Botswana is a middle-income country, 47 percent of its population lives below the national poverty line (UNDP Botswana 2012). As a result, much of the population suffers from the inter-linked phenomenon of persistent poverty and food insecurity (Frayne et al. 2010).

c) Angola and oil

Angola's crude oil production has skyrocketed in recent years. It is now the second largest producer of crude oil in Africa, after Nigeria, and the main supplier to China. As a result, Angola now has the third largest economy in sub-Saharan Africa, behind South Africa and Nigeria, and is perhaps the continent's hottest market at this time because of its oil boom. Between 2004 and 2008, Angola's GDP grew by an average of 17 percent a year (reaching a high of 22 percent in 2007). The economy is expected to grow by 9 percent in 2012 (GGA 2012b).

However, despite this tremendous economic growth, government statistics show that half of the population still lives on less than US$2 a day, and the infant mortality rate remains high: one in five children dies before the age of 5 (GGA 2012b). The cost of living in Luanda, in particular, has skyrocketed with the oil boom and arrival of large numbers of expatriates. In many instances, this has made it difficult for ordinary Angolans to make ends meet.

Not unlike other economies that are dominated by mineral extraction (such as the Botswana case discussed previously), the level of employment created for Angolans by the oil industry is quite limited. As such, Angola's economy needs to be diversified to ensure that a broader segment of Angolan society prospers from this natural resource wealth. To that end, economists at the World Bank and the International Monetary Fund (IMF) have been urging the Angolan government to diversify its economy by establishing a stronger private sector that can create jobs and help distribute the wealth. There have been tepid steps in this direction, with various "investments designed to restore the once-profitable agricultural sector, boost local manufacturing and give small businesses start-up loans" (GGA 2012b: 6). An underlying problem may be that an estimated three quarters of Angola's government revenue is derived from oil sales. As such, the motivation for members of government to diversify the economy may be muted if they are in a good position. This type of conundrum is further explored in the following section on theoretical explanations for why countries get stuck as primary producers.

Summary of the theoretical literature: Do natural resources intrinsically impede economic diversification?

This section briefly reviews the theoretical literature that relates to a fundamental question explored in this chapter, that is, whether or not natural resources intrinsically impede economic diversification. While theory may appear abstract and removed from the practical questions at hand, it is important for understanding the larger reasons for specific phenomena observed on the ground in different African countries. As will be discussed below, some of these explanations or theories focus on the global economic order and trading relationships that exist between countries, whereas others are attuned to the contours of national economies and the nature of natural resources themselves. Here, we do not privilege one set of explanations over another, but merely set them out as potentially useful frameworks to consider. We begin with modernization theory, before turning to dependency and world systems theories, and then eventually Dutch disease and the resource curse.

a) Modernization theory

A suite of theories was developed in the aftermath of World War II to explain how development ought to proceed. These theories, often collectively referred to under the umbrella term "modernization," posited that the industrial economy was the ideal or pinnacle state of development. Theorists of this genre argued that with the right combination of capital, know-how, and attitude, economic growth would proceed. Countries would make a transition from an economy focused on subsistence to one based on primary production for external markets and then to a modern, industrial state.

There were several practical implications of modernization theory for African development policy at the time. In many areas, there was a big push to industrialize agriculture and to build infrastructure such as dams and roads. The idea was that such big infrastructure investments, especially dams, could jump-start an economy and put it on the path to industrialization by providing irrigation for commercial agriculture and cheap electricity for manufacturing. Roads could open up new areas of the country and connect them to the national market economy.

Modernization theory employed a "pull yourself up by the boot-straps" type of philosophy. If a country adopted the right type of policies and made the right type of investments, it would develop. While such autonomy was conceptually appealing because of its tidiness, many argued that it was not the way development occurred in the real world. Multiple critiques of modernization theory emerged in subsequent decades—several of which have been incorporated into what we might loosely call structuralist theories of development.

b) Dependency and world systems theory

In reaction to modernization theory, a new set of development ideas began to arise in the 1960s that emphasized structure, or the global framework under which countries operated. This emphasis on structure meant that relationships between countries were equal to or more important than internal policies for determining the future development of a country.

Dependency theory, originally conceived by Andre Gunder Frank (1979), is one example of such a theory. This approach suggested that economies in the tropics (Africa, Latin America, and Asia) were essentially "underdeveloped" during the colonial era as European countries refashioned these economies (through a combination of taxation policies and coercion) for their own benefit. According to Frank, elites in developing countries often colluded with the colonizers to organize primary extraction in exchange for wealth and power. In many instances, farmers in tropical countries were encouraged or forced to produce commodity crops for the European market (often at the expense of subsistence production) and/or pushed to leave their own farms to pursue wage labor on large plantations or in emerging mining sectors. These primary commodities from the tropics were traded for manufactured goods from Europe and other industrialized countries in a global system of unequal exchange that favored Europeans. Taylor (1992) described the creation of the modern world economy as being "made by Europeans for Europeans as one great functional region centered on Europe."

Wallerstein (1979) expanded on Frank's ideas through his world systems theory, which basically gave a spatial face to dependency theory by depicting the world in terms of the core, semi-periphery, and periphery. Under this schema, the core countries represent the most developed states, which are dominated by industry, financial services, and information-based economies. The semi-periphery is an emerging group of states in which a high volume of cheap manufacturing increasingly takes place. Finally, the periphery represents those countries whose primary role in the global economy is to provide raw materials. All three of these regions operate as a world system, with deep connections existing among each sphere. Most countries in tropical Africa still play the role of the peripheral producers in the global system, with the most advanced African economies (South Africa, Egypt, Tunisia, and Morocco) located in the temperate zone. The problem, as world systems theorists see it, is that many tropical countries will find it difficult to break out of their role of producing goods with peripheral processes, including relatively low wages, low levels of technology, and low levels of education. In other words, Rostow's stages of economic growth will not occur because the least developed countries are locked into an exploitative set of relationships with more developed countries.

While colonialism has long since ended, some would argue that these economic relationships, characterized as dependency, persist in the postcolonial era and inhibit economic diversification beyond primary production. Clearly, African states ought to have the agency or ability to break out of dependency and pursue a path of economic diversification, greater export returns, and higher wage jobs. Some conceptual frameworks may, wittingly or unwittingly, reinforce the status quo. Since the early 1980s, and with the rise of neoliberalism at the world's major financial institutions, the theory of comparative advantage has been used to suggest that countries are better off specializing in a few sectors where they have higher levels of efficiency, special skill sets, and natural endowment. This argument was used in the 1980s and 1990s, backed up by the financial power of the World Bank and the IMF, to double down on primary production in much of tropical Africa. What this argument does not make explicit is its bias toward the status quo by often ignoring the history and policies that went into creating comparative advantage, a situation which is sometimes cast as natural or a fait accompli (Carmody 1998).

c) Dutch disease and resource curse theories

It might seem that an abundance of mineral or energy resources would only bode well for a country. After all, is not growth in countries like the United States, Canada, or Australia at least partially explained by an abundance of natural resources? Some scholars have argued that natural resource wealth actually does more harm than good because groups may fight over control of locally abundant and lucrative resources (such as diamonds in Sierra Leone); it allows governments to become corrupt and detached from the electorate because they are relatively autonomous given rents from resource extraction (as with oil in Libya); and it leads to a stronger currency, which may hurt the country's other exports internationally and facilitate the importation of products that may compete with local industries (as with oil in Nigeria). The first two aspects noted above tend to fall under the natural resource curse problem, whereas the third is known as Dutch disease.

Collier (2007) has asserted that the natural resource curse creates a greater risk of civil war in countries that depend on primary commodity exports. The resource curse may breed corruption among government officials and make them less responsive to public concerns, especially when the government becomes increasingly reliant on revenue from resource extraction (and less dependent on tax revenue). Analysts have suggested that Libya is a good example of a country that suffers from the natural resource curse. In October 2011, Libyan leader Moammar Gadhafi was killed barely clinging to power, trying to stave off rebels (inspired by the Arab Spring of 2011) who controlled large parts of his country. Many experts argued that the structure of Libya's economy allowed Gadhafi to remain in power for 41 years. Libya's economy was based almost entirely on resource exports, with 98 percent related to oil and gas (NPR 2011). Gadhafi's government made money by selling oil to the rest of the world, rather than collecting taxes. It used these funds to buy guns, mercenaries, and the loyalty of some citizens. Gadhafi also dissuaded Libyans from starting their own businesses, which would be a source of revenue he could not control, and a potential rival source of power. But the resource curse is not inevitable, and Libya's next leader will not necessarily take over the oil wells and become another dictator. Libya's wealth from fossil fuels could also provide it with the opportunity to explore new forms of government.

Watts (2004) has been critical of the natural resource curse thesis on the grounds that conflicts over resources are often a symptom, rather than a cause, of underlying tensions. Furthermore, to imbue a set of natural resources with the power to cause problems is to believe that the resource alone has a special transformative power. Some commentators have argued that resource curse cases involving fossil fuels may deserve special consideration, given the dependence of the outside world on these energy resources (Moseley 2009). Energy-hungry countries (such as the United States and China) are often willing to support undemocratic and corrupt regimes if it means they will have reliable access to oil or natural gas. In these instances, oil's use in the larger global economy helps to create the unhealthy dynamic inside exporting countries.

In tandem with the natural resource curse, overreliance on a small set of commodity exports may lead to the simplification of a national economy (Gylfason 2001). Known as Dutch disease, the problem gets its name from Holland, because the Dutch economy suffered from such a simplification in the 1960s after natural gas was discovered in the North Sea. As exports of a commodity increase, so do inflows of foreign exchange, which raise the value of a nation's currency. Dutch disease occurs when a more highly valued currency makes a nation's other exports more expensive and less competitive in the global market. Furthermore, a higher-valued national currency also makes imports relatively inexpensive, and these cheap imports pose problems for domestic producers. Both of these trends (relatively more expensive exports and cheaper imports) tend to hurt a country's local industries.

Nigeria in particular is thought to have suffered from Dutch disease (Olusi and Olagunju 2005). Prior to the development of the oil industry, Nigeria had relatively robust agricultural and manufacturing sectors that served the rest of West Africa. However, as the oil industry grew, the manufacturing industry and agriculture declined precipitously. Even though corruption was and is also an issue in Nigeria, which has consistently ranked among the most corrupt countries in the world over the past several years (Transparency International 2011), Dutch disease has also been a factor.

Approaches to economic diversification in resource-rich African countries

a) The rise, fall, and rebirth(?) of import substitution

Dependency theory and world systems theory led to some real policy changes on the ground in African countries during the 1960s and 1970s. Probably the most significant of these was an approach known as "import substitution," often associated with the Argentine economist Raúl Prebisch. Prebisch (1959) argued that tropical countries would be forever stuck as producers of primary products and therefore would fail to develop unless they took proactive steps to change the nature of their economies in relation to those of others. The idea behind import substitution was that manufactured goods needed to be produced at home rather than imported from the core countries. Given a lack of private capital available for industrialization and stiff competition from producers in the core countries, many African governments became directly involved in the creation of such enterprises. Such state-run enterprises came to be known as parastatals. Governments typically also erected tariff barriers to protect such nascent or infant industries until they could stand up to international competition.

Import substitution was quite popular until the late 1970s when the so-called Third World debt crisis struck. This crisis involved a number of developing countries, including several in Africa, that were on the verge of defaulting on loans owed to creditors (largely public creditors in the case of Africa). The crisis was brought on largely by government involvement with increasingly inefficient state-run enterprises and the energy crisis of the 1970s, when high oil prices were a challenge for many African oil importers.

From the 1980s through the 2000s, public lenders responded to the debt crisis by imposing a new neoliberal economic order that stressed small government, free trade, export orientation, and the privatization of state-run enterprises. Furthermore, the neoliberal era brought a decisive end to two decades of experimentation with import substitution in the African context. Critics of neoliberalism suggested that this was a return to the economic policies of the colonial era. Furthermore, there was a concern that there had been a selective reading of history in order to make the case that import substitution was overly problematic and that a narrow focus on comparative advantage and commodity exports was the cure (Carmody 1998). With the NICs of the Asian Pacific Rim being held up as the new development models, the storytellers conveniently ignored the active role that the state had played in initially protecting and nurturing what would become their major export industries. More recently, some economists have begun to argue that import substitution should be revisited in some cases as an economic policy (Brutton 1998; Amsden 2003).

b) New structural economics

A more recent development is "new structural economics," an approach closely associated with the former chief economist of the World Bank, Justin Lin (2012). According to Lin, the market is the most effective mechanism for resource allocation at each level of development. However, economic development necessitates that economies evolve (by moving from primary to secondary to tertiary economic activities) and become more diversified. Furthermore, this evolution is often heavily dependent on improvements in what he calls hard and soft infrastructure, or the physical infrastructure and know-how needed to promote certain types of economic activities. Because such improvements in hard and soft infrastructure often require coordination and are a form of public good (a good needed by all, but which no individual firm would be willing to develop), government should play an active role in developing such infrastructure. Lin also draws on the ideas of the Japanese economist Kaname Akamatsu (1962) and his "flying geese paradigm" to argue for regional economic integration between different types of economies. Writing from the perspective of Japan in the 1960s, Akamatsu argued that Asian nations would catch up with the West by forming an Asian regional block within which the poorest countries produced the commodities that were subsequently transformed in the more advanced countries. The metaphor was a flock of geese in V formation, with Japan in the lead, then the second tier of NICs (South Korea, Taiwan, Singapore, and Hong Kong), followed by the lesser-developed countries in the region.

The new structural economics approach holds some ideas in common with import substitution in the sense that it envisions a role for the state in creating new industries. However, the approaches are different in terms of where one might invest resources: import substitution favors direct support to industries (often via protective tariffs or subsidies), while structural economics emphasizes investment in the physical infrastructure and know-how that would facilitate the rise of a new industry.

c) Trust fund approach

A third and final tactic is the so-called trust fund approach, wherein one effectively takes a nonrenewable resource and converts it to a renewable resource. This is done by reinvesting the proceeds from resource extraction into a trust fund that yields an income flow over time (El Serafy 1991). A side benefit of this approach is that it has the potential to remove the temptation for corruption by taking the resources out of state civil servants' hands, a critical factor identified by some scholars (Gelb 2011). The trust fund approach has been tried in a few countries, both less and more developed. A good example of a less developed nation that has pursued this approach is the Micronesian island state of Kiribati, which has a population of 90,000 people spread over 34 islands and a per capita income (2004) of US$950. While most of the population (80 percent) is engaged in a subsistence-based farming and fishing economy, the country also has significant phosphate deposits, which it is mining and exporting. Since 1956, the proceeds from phosphate extraction have been placed in a trust fund that is invested offshore by two London-based account managers (Gibson-Graham 2004). The returns on this fund are used to finance government services, including health care and the development of a communication and transportation infrastructure among the islands. This means that most residents in Kiribati are free to continue living a subsistence lifestyle, yet still have access to sustainably financed government services.

An example from a developed country is found in Norway, where proceeds from the natural gas industry have been placed in a trust fund that subsidizes government service provision (Bantekas 2005).

Both cases represent situations where nonrenewable resources (phosphate and natural gas) have effectively been converted to a renewable resource (a self-sustaining trust fund) that provides for ongoing investments in a country's human capital.

Botswana and "on-shoring" of value-added activities to the diamond industry

As discussed earlier, while Botswana is considered to be a highly successful African economy (Samatar 1999) and one of the few to move from being a low-income to a middle-income country, it still has a relatively undiversified economy that is highly dependent on diamonds. Despite Botswana's relative prosperity, there is recognition that its overdependence on diamonds is a problem. Most recently, there has been a drop in global diamond demand as a result of the international recession, with a subsequent contraction of Botswana's diamond export revenue.

As such, the Government of Botswana has made it an explicit policy objective to try to diversify the economy. It has set about this in a couple of different ways. First, the government is attempting to foster more value-added processing of diamonds within the country. Until recently, most of the country's precious diamonds had been exported in rough form to other locations to be cut and polished, so that the country missed out on critical value-added economic opportunities. The goal of the "diamond beneficiation program" is to maximize the economic benefits of diamond extraction in Botswana (Leach 2011). Tiffany & Co. has recently established Laurelton Diamonds, a local subsidiary, in Botswana to process its diamonds within the country. It is estimated that the partial on-shoring of the diamond finishing business in Botswana will bring in an extra US$8 million in revenue to the country. At the time of writing, Botswana's capital, Gaborone, has been in the process of receiving skilled laborers from around the world with experience in diamond cutting and finishing. The plan is for these skills to eventually be transferred to Botswana's own population. While it is still too early to make a pronouncement on the success of this state-led initiative, the early signs look promising.

The Government of Botswana's other attempt to promote economic diversification has been via efforts to encourage entrepreneurialism in the country. In contrast to the previous program described, the results for this initiative have been less promising. The Government of Botswana has launched a number of programs to encourage entrepreneurialism. These initiatives have included training for entrepreneurs through the University of Botswana as well as loans for business proposals (via the Citizen Entrepreneurial Development Agency). To date, the repayment rate on the loan program has been low and the number of successful new businesses launched has not been high. Some argue that there is an inadequate entrepreneurial culture in Botswana. The country's lack of private sector entrepreneurial activity is in part historical, as this was actively discouraged during the colonial period, when only Europeans and Asians were granted commercial licenses (Samatar 1999). Others suggest that a civil service position remains the ultimate goal in Botswana society, and entrepreneurial activity is perceived as less prestigious, although an important secondary activity (Chart 2012).

An important lesson to be taken from the Botswana experience is that the state is taking an active role in developing the soft infrastructure or know-how needed to support new forms of economic activity. In building a value chain based on the diamond industry as well as a more entrepreneurial culture, the state has seen the need to provide training via a university or by fostering the transfer of knowledge between experienced diamond cutters and those entering this highly skilled arena

The way forward: Policy recommendations for diversification beyond primary production

Today, many African countries continue to ride a natural resource and commodity export boom. Furthermore, many experts believe that this expansion is driven by a set of demographic and growth pattern factors that make it distinct from past booms, leading some to conclude that a continued African focus on primary production is the continent's best bet for sustained prosperity. What such thinking ignores is the very nature of commodity markets, which, in the absence of effective cartels, have experienced price declines over the long term because of relatively low barriers to entry for some types of commodity production (compared with the secondary and tertiary production sectors), price thresholds that trigger the development and expansion of substitutes, and technological advances that expand the size of exploitable reserves. In addition, undiversified, commodity-based economies are highly vulnerable to increasingly volatile global markets.

African economies must diversify away from primary production (resource extraction and agricultural commodity production) if they are to improve their economic position vis-à-vis the rest of the world. Some African economies have begun to diversify by developing more extensive value chains in existing extractive industries, whereas others have seized on new windows of opportunity. In both instances, we are seeing the beginnings of effective structural transformation.

A review of development theories and case studies suggests that a number of steps should be explored in order to transform and diversify resource-based African economies.

First, while the market may be an efficient allocator of resources at any particular stage of economic growth, development necessitates a diversification of the economy and evolution from primary to secondary to tertiary forms of economic activity. As new forms of economic activity are often dependent on certain forms of physical infrastructure and know-how, the state has a key role to play in creating such conditions. As such, African governments need to think critically about strategic infrastructure and education investments.

Second, there is great potential for building value chains based on existing forms of commodity production. Often this means transforming a raw product into something more valuable for export by identifying potential areas for value chain development, encouraging investment along the value chain, and supporting the development of the necessary soft and hard infrastructure.

Third, in some cases, it may make sense for poor African countries with recently discovered natural resource wealth to develop a trust fund approach that converts a nonrenewable resource to a renewable resource that generates a steady flow of income over time. This income stream should be used for critical investments in human capital (education and health care) that create the possibility for future economic diversification.

Finally, encouraging regional economic integration, as per the flying geese paradigm, may make sense, as it leads to the development of industrial activity in a subregion of the continent and stems the outflow of talent to other regions of the world. These could eventually lead to affiliated economic activity in other areas of the region. The development of such integration hinges on relaxing tariff barriers among countries in the subregion where there are potential economic complementarities (such as between coastal West Africa and interior countries).

Notes

1. This boom continued for all but a few luxury items such as diamonds (of critical import to African exporters such as Botswana, South Africa, and Congo-Brazzaville).
2. However, Tanzania and Mozambique now have recently discovered natural gas resources, which may change their classifications.

References

African Capacity Building Foundation (ACBF). (2012). *Africa Capacity Indicators 2012: Capacity Development for Agricultural Transformation and Food Security*, Harare, Zimbabwe: ACBF.

African Development Bank Group. (2012). "Overview of the Economic Situation and the Role of the Bank," in *2010 Annual Report*, Abidjan, Côte d'Ivoire: African Development Bank Group: Chapter 1.

Akamatsu, K. (1962). "A Historical Pattern of Economic Growth in Developing Countries," *Journal of Developing Economies*, 1(1): 3–25.

Amsden, A. (2003). *The Rise of the Rest: Challenges to the West from Late Industrializing Economies*, London: Oxford University Press.

Baden, S. and Barber, C. (2005). *The Impact of the Second-Hand Clothing Trade on Developing Countries*, Available at: http://rkr6.irec.no/dm_documents/Oxfams%20rapport%202005_V8dcF.pdf. Accessed March 1, 2013.

Bantekas, I. (2005). "Natural Resource Revenue Sharing Schemes (Trust Funds) in International Law," *Netherlands International Law Review*, 52: 31–56.

Bassett, T.J. (2010). "Slim Pickings: Fairtrade Cotton in West Africa," *Geoforum*, 41(1): 44–55.

Bergmann, L. (2012). "Beyond Imagining Local Causes/Solutions to a Global Problem: Mapping Carbon Footprints of Global Capitalism," Annual Meeting of the Association of American Geographers, New York, NY.

Bruton, H.J. (1998). "A Reconsideration of Import Substitution," *Journal of Economic Literature*, 36: 903–936.

Carmody, P. (1998). "Constructing Alternatives to Structural Adjustment in Africa," *Review of African Political Economy*, 25(75): 25–46.

Carmody, P. (2011). *The New Scramble for Africa*, Malden, MA: Polity Press.

Carmody, P. and Owusu, F. (2007). "Competing Hegemons: Chinese vs. American Geo-Economic Strategies in Africa," *Political Geography*, 26(5): 504–524.

Carmody, P. and Taylor, S. (2003). "Industry and the Urban Sector in Zimbabwe's Political Economy," *African Studies Quarterly*, 7(2/3): 53–80.

Central Intelligence Agency (CIA). (2012). *The World Factbook*, https://www.cia.gov/library/publications/the-world-factbook/. Accessed March 1, 2013.

Chart, H. (2012). "Proof of Passion: Struggles to Identify and Become 'Real Entrepreneurs' In Urban Botswana," Annual Meeting of the American Anthropological Association, November 14, San Francisco, CA.

Collier, P. (2007). *The Bottom Billion*, New York: Oxford University Press.

Elliot, L., Sieper, H. and Ekpott, N. (2011). *Redefining Business in the New Africa: Shifting Strategies to Be Successful*, Charlotte, NC: Conceptualee, Inc.

El Serafy, S. (1991). "The Environment as Capital," in R. Constanza (ed.), *Ecological Economics: The Science and Management of Sustainability*, New York: Columbia Press.

FAO. (2006). FAOstat (an online database of the Food and Agricultural Organization of the United Nations), http://faostat.fao.org. Accessed March 1, 2013.

Frank, A.G. (1979). *Dependent Accumulation and Underdevelopment*, New York: Monthly Review Press.

Frayne, B., Pendleton, W., Crush, J., Acquah, B., Battersby-Lennard, J., Bras, E., et al. (2010). *The State of Urban Food Insecurity in Southern Africa*, Urban Food Security Series No. 2, Kingston, ON, and Cape Town: Queen's University and AFSUN.

Gelb, A. (2010). *Economic Diversification in Resource Rich Countries*, Washington, DC: International Monetary Fund.

Gelb, A. (2011). "Natural Resource Exports and African Development," in E. Aryeetey, S. Devarajan, R. Kanbur, and L. Kasekende (eds.), *Oxford Companion to Economics in Africa*, New York: Oxford University Press.

Gibson-Graham, J.K. (2004). "Surplus Possibilities: Re-Presenting Development and Post-Development," *Conference on Economic Representations: Academic and Everyday*, University of California Riverside. April.

Good Governance in Africa (GGA). (2012a). "Are Diamonds Mugabe's Best Friend?" *Africa in Fact*, 3(August): 25–28.

Good Governance in Africa (GGA). (2012b). "Angola's Oil Riches Stream into the Pockets of an Entrenched Elite," *Africa in Fact*, 3(August): 5–8.

Government of Mali (GoM). (1998). *Plan National d'Action Environnementale*, Bamako, Mali: Ministry of Environment.

Gylfason, T. (2001). "Natural Resources, Education, and Economic Development," *European Economic Review*, 45(4–6): 847–859.

Hillbom, E. (2008). "Diamonds or Development? A Structural Assessment of Botswana's Forty Years of Success," *Journal of Modern African Studies*, 46: 191–214.

Horta, L. (2009). "Food Security in Africa: China's New Rice Bowl," *China Brief*, 9/11 (May): 9.

Kaberuka, D. (2007). "Managing Africa's Commodity Boom" (interview with Dr. Donald Kaberuka, President of the African Development Bank Group), December 14, http://www.royalafricansociety.org/the-africa-business-breakfasts/447.html. Accessed October 15, 2012.

Leach, A. (2011). "Diamond Finishing to Boost Botswana Economy by $800 Million," *Supply Management*, November 25, http://www.supplymanagement.com/2011/diamond-finishing-to-boost-botswana-economy-by-800million/. Accessed March 1, 2013.

Ledermann, S.T. and Moseley, W.G. (2008). "The World Trade Organization's Doha Round and Cotton: Continued Peripheral Status or a 'Historical Breakthrough' for African Farmers?" *African Geographical Review*, 26: 37–58.

Leke, A., Lund, S., Roxburgh, C., and van Wamelen, A. (2010). "What's Driving Africa's Growth?" *McKinsey Quarterly* (June), http://www.mckinseyquarterly.com/Whats_driving_Africas_growth_2601. Accessed March 1, 2013.

Leppman, E. (2005). *Changing Rice Bowl: Economic Development and Diet in China*, Hong Kong: Hong Kong University Press.

Lin, J. (2012). *New Structural Economics: A Framework for Rethinking Development and Policy*, Washington, DC: World Bank.

Moseley, W.G. (1993). "Indigenous Agroecological Knowledge among the Bambara of Djitoumou Mali: Foundation for a Sustainable Community," master's thesis, University of Michigan, Ann Arbor.

Moseley, W.G. (2009). "'Response to Michael Watts'. Whither Development?: The Struggle for Livelihood in the Time of Globalization," *Macalester International*, 24: 140–149.

Moseley, W.G. (2012) "Assaulting Tolerance in Mali," *Al Jazeera—English*, July 16 (print), http://www.aljazeera.com/indepth/opinion/2012/07/20127159 4012144369.html. Accessed March 1, 2013.

Moseley, W.G., Carney, J., and Becker, L. (2010). "Neoliberal Policy, Rural Livelihoods and Urban Food Security in West Africa: A Comparative Study of the Gambia, Côte d'Ivoire and Mali," *Proceedings of the National Academy of Sciences of the United States of America*, 107(13): 5774–5779.

Moseley, W.G. and Gray, L.C. (eds.) (2008). *Hanging by a Thread: Cotton, Globalization and Poverty in Africa*, Athens, OH: Ohio University Press and Nordic Africa Press.

Muldavin, J. (2007). "China's Not Alone in Environmental Crisis," *Boston Globe*, December 19.

National Public Radio (NPR). (2011). "Economists Diagnose Libya with 'Resource Curse,'" February 25, http://www.npr.org/2011/02/25/134048260/ Libyas-Economy. Accessed March 1, 2013.

Naude, W., Bosker, M., and Matthee, M. (2010). "Export Specialisation and Local Economic Growth," *World Economy*, 33(4): 552–572.

Olusi, J.O. and Olagunju, M.A. (2005). "The Primary Sectors of the Economy and the Dutch Disease in Nigeria," *Pakistan Development Review*, 44(2): 159–175.

Oramah, B.O. (2012). "From Commodities to Raw Materials: Afreximbank's Approach to Promoting Commodity Transformation in Africa," *Global Commodities Forum*, Geneva, January 23–24.

Oxfam (2003). *Cultivating Poverty*, Oxfam Briefing Paper no. 30, London: Oxfam UK.

Prebisch, R. (1959). "Commercial Policy in the Underdeveloped Countries," *American Economic Review*, 49: 251–273.

Roberts, R.L. (1996). *Two Worlds of Cotton: Colonialism and the Regional Economy in the French Soudan, 1800–1946*, Palo Alto, CA: Stanford University Press.

Roxburgh, C., Dörr, N., Leke, A., Tazi-Riffi, A., van Wamelen, A., Lund, S., et al. (2010). *Lions on the Move: The Progress and Potential of African Economies*, Washington, DC: McKinsey Global Institute.

Samatar, A. (1999). *An African Miracle: State and Class Leadership and Colonial Legacy in Botswana Development*, Portsmouth, NH: Heinneman.

Sharma, R. (2012). "We Should Celebrate the End of the Commodity Supercycle," *Financial Times*, June 24.

Taylor, P.J. (1992). "Understanding Global Inequalities: A World-Systems Approach," *Geography*, 77(1): 10–21.

Tefft, J. (2004). "Mali's White Revolution: Smallholder Cotton from 1960 to 2003," *Building on Successes in African Agriculture*, Focus 12, Brief 5, Washington, DC: International Food Policy Research Institute.

Transparency International. (2011). "Corruptions Perception Index 2011," http://cpi.transparency.org/cpi2011/results/. Accessed March 1, 2013.

Tverberg. G. (2012). "Why US Natural Gas Prices Are So Low—Are Changes Needed?" *Our Finite World*, March 23, ourfiniteworld.com. Accessed March 1, 2013

United Nations (UN). (2010). "UN Data Statistics," Available at: http://data.un.org/Data.aspx?d=SNAAMA&f=grID%3A101%3BcurrID%3AUSD%3BpcFlag%3A1. Accessed March 1, 2013.

United Nations Development Programme (UNDP) Botswana. (2012). "Poverty Reduction," http://www.unbotswana.org.bw/undp/poverty.html. Accessed March 1, 2013.

United States Department of Agriculture (USDA). (2012). "Coffee: World Markets and Trade," June, Washington, DC: USDA Foreign Agricultural Service, http://www.fas.usda.gov/psdonline/circulars/coffee.pdf. Accessed March 1, 2013.

Wallerstein, I.M. (1979). *The Capitalist World Economy: Essays*, New York: Cambridge University Press.

Watts, M. (2004). "Resource Curse? Governmentality, Oil and Power in the Niger Delta, Nigeria," *Geopolitics*, 9(1): 50–80.

6

Strategic Capacity-Building Imperatives Vital for Transboundary Water Cooperation in Africa

Claudious Chikozho

Introduction

Several of the 55 transboundary river basins in Africa are shared by ten or more countries and this makes transboundary river basin management (TRBM) in the continent quite complex (Sadoff et al. 2002; UNECA 2006). But the arbitrary subdivision of landscapes into new countries during colonialism and the high variability of rainfall patterns and river flows across the continent make TRBM challenges in Africa even more complex (SIWI 2009). Despite attempts by various scholars and practitioners in recent decades to articulate the main drivers and constraints to TRBM, knowledge and understanding about the key factors that determine successful interstate cooperation and capacity-building interventions in this domain remains limited. Thus, the search for effective TRBM and proactive solutions for addressing interstate water conflict remains an ongoing effort in Africa (Sadoff and Grey 2002; GWP 2012).

While Africa's transboundary water resources are less developed than those in other parts of the world, recurring cycles of droughts and floods accentuate water scarcity across the continent (Falkenmark 1989; Chenje 1996; Rieu-Clarke et al. 2012). At the same time, most of the dominant projections of Africa's water profile and climate change patterns in the coming decades paint a picture of worsening scarcity in various "hot spots" such as the Nile, the Niger, Zambezi, and Okavango river basins (Chenje 1996; FAO 2004; IFPRI 2007). Large storage reservoirs are certainly required to store water for irrigation, hydropower, and reducing impacts of annual precipitation variations (Merrey 2009). Such large projects have serious interstate cooperation implications since, in many cases, their construction is only feasible on rivers shared by two or more countries.

A key objective in TRBM is to find ways of turning what is often perceived as a zero-sum predicament—in which one party's gain is another's loss—into a win-win proposition (Zaag and Savenije 2001). Indeed, there is now a well-articulated need to systematically build capacity at various levels and enhance the ability of key actors to lead and/or facilitate collaborative TRBM. Using evidence from selected case studies and the published literature, this chapter assesses TRBM regimes, highlights main challenges and opportunities, and generates specific recommendations that may inform future capacity-building interventions in this landscape (Map 6.1).

Conceptualizing capacity building

As an analytical construct, the concept of "capacity building" still lacks a universally shared definition. Fully articulated frameworks for assessing capacity needs, designing and sequencing appropriate interventions, and determining results are also lacking (Boesen and Ravnborg 2004; World Bank 2005). Traditional approaches to capacity building tend to narrowly focus on strengthening the skills of individual officials. But emerging approaches emphasize addressing capacity at the levels of individuals, organizations, institutions, policies, and the whole society (Fukuda-Parr 2003). When viewed in this way, capacity building entails improving the set of systemic attributes and human capital resources that enable individuals, organizations, and societies to sustainably articulate and actualize their developmental goals, building on their own resources (Saasa 2007; AU/NEPAD 2009). For the purposes of this chapter, "capacity building" is defined as the process through which individuals, organizations, and societies in TRBM contexts obtain, strengthen, and maintain the capabilities to set and achieve their own basin development objectives (UNDP 2008; UN-Water 2009).

In a review of 40 years of development experience, the Organisation for Economic Co-operation and Development (OECD) concluded that donors and partner countries alike have tended to look at capacity building as mainly a technical knowledge transfer process from developed to developing countries (OECD-DAC 2006). The African Union-New Partnership for African Development AU-NEPAD Capacity Development Initiative calls for broader and deeper analysis of capacity needs, better

Map 6.1   Main transboundary river basins in Africa
Note: Product of the Transboundary Freshwater Dispute Database, Department of Geosciences, Oregon State University. Additional information about the TFDD can be found at: http://www.transboundarywaters.orst.edu.

understanding of prevailing deficiencies, and deployment of innovations to address the deficiencies (AU/NEPAD 2009). From this perspective, capacity building also encompasses an analysis of the system as a whole in order to develop more sustainable solutions.

Paradigmatic shifts in water sharing

A relatively vociferous group of scholars has emerged arguing that conflicts over natural resources increase and escalate in tandem with worsening resource scarcity (Sadoff and Grey 2005; Wirkus and Böge 2006; SIWI 2009). Thus, increasing water scarcity in Africa in the face of rising demand is viewed as creating fertile grounds for armed conflict, both within and between countries (Ashton 2000; Biswas and Tortajada 2009). In this context, the water conflict challenge lies at the nexus between what is essentially a "zero-sum game" and the peculiarities of the upstream-downstream dilemma, in which water withdrawals and use by one upstream state reduce the quantity and quality of water available to the other riparians downstream (Zaag and Savenije 2001; UNECA 2006).

While the voice of the proponents of the so-called water-wars thesis might have sounded loudest in the last two to three decades, there are a substantial number of scholars who argue that the predicted violent international conflicts are not only unnecessarily alarmist, but are also not based on solid empirical evidence. They further argue that one can hardly find historical cases where countries actually engaged in violent conflict solely due to disagreements over transboundary water sharing. It is usually factors outside the water domain that are decisive in exacerbating transboundary tensions and not water per se (Uitto and Duda 2002; Jacobs 2009; Douven et al. 2012). Since 1948, only 37 incidents of acute conflict over water have been witnessed compared to approximately 295 international agreements for cooperation over water that were negotiated and signed during the same period (Wolf 2005; Conca 2006). Riparians have actually shown tremendous creativity by moving thinking beyond a single focus on polarized claims for water sharing in volumetric terms to initiation of joint regional development projects and creation of broader "baskets of benefits" which allow for positive-sum allocations of joint gains (UN-Water 2008).

The number of international water cooperation arrangements and Transboundary River Basin Organizations (TRBOs) in Africa continues to increase, and efforts in that direction should be further promoted. But several analyses suggest that only a few African Regional Economic Communities have been able to actually develop meaningful regional water policy and strategic action plans for integrated water resources management and facilitate implementation of some regional projects (Wirkus and Böge 2006; UNECA 2006; Jacobs 2009). Practically establishing workable agreements on water sharing and facilitating greater regional economic integration remain enormous challenges in the continent and beyond (Gerlak 2007; Merrey 2009).

The case studies

The challenges and opportunities evident in the case studies presented as follows highlight the basic resource management and capacity-building priority issues that arise in transboundary water basins.

The Niger Basin Authority

The Niger River basin covers more than 2 million km^2 and is shared by 11 countries, namely, Algeria, Benin, Burkina Faso, Cameroon, Chad, Côte d'Ivoire, Guinea, Sierra Leone, Mali, Niger, and Nigeria (Bach et al. 2012). The Niger Basin Authority (NBA) was established to undertake a number of activities, including the gathering, standardization, and dissemination of data, the development of joint plans for infrastructure development and transport, the establishment of norms and activities for preventing and reducing environmental threats, especially in the field of water pollution, and the promotion of agriculture, forestry, and fisheries (AMCOW 2007). The NBA also seeks to harmonize and coordinate the national water policies of the member states.

Some of the detailed operational objectives already attained include establishment of a sustainable basin development action plan derived from a collaborative decision-making process, establishment of an institutional framework that lays the foundation for dialogue among the riparian countries, and adoption of a consensus-building approach in the development of the shared vision (Burchi and Spreij 2003). Three dams have been commissioned for construction to provide for hydroelectricity, irrigation, and navigation in the basin. Anticipated outcomes of constructing these dams include a fivefold increase in irrigated agriculture by 2027 and creation of many jobs (Bach et al. 2012). The authority has conducted several projects to tackle specific problems in the basin such as aquatic weed control, desertification control, and the promotion of biomass gas production and use. One of its major projects, entitled HYDRONIGER, is designed to establish an operational hydrological forecasting system, assist the member states in their drought and flood control activities, and provide data for agriculture, hydropower, navigation, and other development activities in the basin (NBA/GEF 2002; Bach et al. 2012).

Each member state contributes to the budget of the NBA on the basis of a sharing formula that was agreed upon by the states. The contribution to the budget ranges from 30 percent for Nigeria to 1 percent for Chad (NBA 2004). The basin and the NBA face various challenges. These include land degradation and desertification, water resources reduction, unsustainable exploitation of natural resources, limited institutional capacity, limited stakeholder involvement, and funding (NBA/GEF 2002). Historically, one of the key factors contributing to the shortage of finances is that some of the member countries have often failed to remain up to date with their financial contributions.

Lake Chad Basin Commission

Established in 1964, the Lake Chad Basin Commission (LCBC) is constituted by five riparians to the lake, namely, Cameroon, Niger, Nigeria, Chad, and the Central African Republic (CAR), as well as Sudan, which has observer status. The main functions of the LCBC include preparing joint rules and ensuring their effective application, collecting and disseminating information on projects prepared by member states, facilitating joint planning and research programs within the basin, facilitating efficient water use, promoting and coordinating regional cooperation projects, and examining complaints and resolving disputes (UNECA 2000; Burchi and Spreij 2003).

To date, the main results of the LCBC's work include the establishment of a master plan for the development and environmentally sound management of the natural resources of Lake Chad, which was ratified in 1994 (UNECA 2000; Odada et al. 2004). It should also be noted that the existence of the LCBC for several decades now in a region that has almost constantly been the scene of civil and international strife and other violent conflicts is positive and has ensured that communication over water between the member countries does not break down—even in times of severe crisis (Scheumann and Neubert 2006). The member countries make contributions to the commission's funding based on an agreed-upon formula, which requires Nigeria to pay 52 percent, Cameroon 26 percent, Chad 11 percent, Niger 7 percent, and CAR 4 percent of the commission's annual budget (UNECA 2000). Numerous donors and technical partners and international NGOs have provided funding and technical support for the collection, processing, and exchange of data as well as reversal of land and water degradation trends in the lake's ecosystem (Scheumann and Neubert 2006).

Poor coordination and lack of broad-based stakeholder participation are perhaps the most critical managerial problems confronting the basin. Initially, the member countries committed themselves to a shared use of the basin's natural resources and refrain from implementing, without consulting the commission, any measures likely to significantly affect water availability and ecosystem health (LCBC Convention and Statutes 1964: Article 5). In practice, however, the member states have violated these commitments on several occasions, constructing dams and irrigation projects without prior notification (AMCOW 2007). Consequently, economic development in the basin involves national schemes that are conceptualized in isolation, and individual member states still pursue their own water policy, largely without informing the other members or the LCBC (Odada et al. 2004). The lake is also massively impacted by environmental degradation and existing plans for over-dimensioned, ecologically doubtful, and unsustainable projects provide good reason to anticipate major problems in the future (Isiorho et al. 2000; Bach et al. 2012). Recent efforts to strengthen and improve the commission and lake basin management in general are certainly a good sign, although the future prospects of these efforts are still uncertain (Bach et al. 2012).

The Okavango River basin

The Okavango River basin extends through Angola, Namibia, and Botswana. Namibia and Botswana are two of the driest countries in Southern Africa, and the Okavango River plays an important role not only in the lives of local populations residing along the river, but also at national levels (Klaphake and Scheumann 2006). The Permanent Okavango River Basin Water Commission (OKACOM) was established in 1994 and given the main role of anticipating and reducing unintended impacts that occur due to uncoordinated water resource development (Heyns 2002; Turton 2004). It advises its member states on measures and arrangements to determine the long-term safe yield of water available from all sources in the basin and anticipate demand. It sets criteria to be adopted for the equitable allocation, conservation, and sustainable utilization of water resources in the basin; investigates the construction, operation, and maintenance of any waterworks in the basin; and controls water pollution and aquatic weeds in the basin. OKACOM also facilitates ongoing dialogue among the basin stakeholders to address topical issues (AMCOW 2007).

The commission carried out a basin-wide transboundary diagnostic assessment to identify key areas of concern and gaps in knowledge regarding the physical and socioeconomic system of the basin, with the full participation of key stakeholders (UNECA 2000). The agreement bringing together the three countries stipulates that the riparians should notify the commission and each other of any proposed development (Turton 2004). A number of donors and international NGOs have provided financial and technical support to enable OKACOM to flourish. The member states also share equally some of the costs of running the commission (Heyns 2002).

Technical cooperation in the Okavango basin faces a typical upstream-downstream challenge because water use in one country could affect water availability to the other riparians (Turton et al. 2003). Out of the three riparian countries, Botswana is the most vulnerable to upstream uses and heavily relies on transboundary cooperation to avert crises (Heyns 2002). No specific provisions have been made so far for local stakeholder participation. Other key challenges include lack of basin-wide organizations for water resources management, inadequate political dialogue on integrated water resources management (IWRM), lack of knowledge regarding water demands in Angola, limited capacity for IWRM, high variability of available water resources, and lack of pertinent hydrological data (Ashton 2003; AMCOW 2007).

International experiences

There are a number of lessons emerging from the African case studies, and these also resonate very well with TRBM and capacity-building experiences of transboundary river basins with a long history of water cooperation in other parts of the world. Notable examples include the Mekong and Danube River basins. Lessons of experience from these well-established basins could be used to strengthen similar efforts in Africa. For example, since 1992, governments in the Mekong River basin have been pursuing economic linkages, connecting regional infrastructure, and promoting cross-border trade and collaborative responses to social and environmental problems (Bach et al. 2011; Hall and Bouapao 2011). Despite many political challenges and drawbacks, significant regional economic growth has already been witnessed, twice that of the world's average during 1999–2008 (CIE 2010). The need to strengthen human and institutional capacity has been well articulated and programs developed to address these needs. Various donors have been providing technical and financial assistance to support these efforts (MacQuarrie et al. 2008). Regional training and mapping of hydrological "hot spots" are some of the examples of collaborative interstate cooperation in the basin (Sadoff and Grey 2002).

In the Danube River basin, profound interstate cooperation arrangements have dominated water management for many decades. The European Union, the Global Environmental Facility (GEF), and the United Nations Development Programme (UNDP) have provided substantial funding for capacity development in the basin, and the results of these efforts have been quite positive (ICPDR 2009; Bach et al. 2012). Since its establishment in 1998, the Danube Basin Commission has grown into one of the largest and most active TRBM bodies in Europe and beyond. Some of its key result areas and achievements include ensuring sustainable water management, ecosystem conservation, improvement and efficient use of surface water and groundwater, and controlling water pollution and floods (ICPDR 2009). International expert groups have been set up in the basin to develop strategies and guidelines for more effective resource management, close cooperation, and efficient information exchange among the basin countries (UNOPS 2001).

Lessons from the case studies

The experiences of all the case study basins presented show that the quest for regional economic growth and minimization of interstate conflicts tops the agenda in international water cooperation. In that respect, the primary objective behind the establishment of TRBOs becomes the optimal utilization of the riparian states' collective capabilities and realization of common interests revolving around water use and related resource sectors. A key lesson for capacity-building interventions is that given the wide range of functions that the TRBOs and inter-state commissions assume, an equally wide range of technical assistance to the TRBOs is required. Data gathering, information sharing, hydrological modeling, and development of more robust decision support systems are consistently highlighted as key factors for successful TRBM. Activities to maintain ecological sustainability of the river system are also quite prominent across all the case studies. Therefore, targeted capacity building may be applied as an instrument for improving the utilization of transboundary waters, maintaining ecosystem health, and enhancing regional economic integration (Falkenmark et al. 2009). Failure to embrace and facilitate more broad-based participation of transboundary basin stakeholders stands out as a serious blemish to the history of the African case studies presented.

The demand for capacity building in TRBM is high, and this is also reflected in the large number of internationally funded projects and technical support provided to the transboundary basin states. For example, the Food and Agricultural Organization, the Global Environmental Facility, German Technical Cooperation (BMZ), and the UNDP have been helping many transboundary basin countries to establish legal and institutional environments conducive to stable and mutually beneficial interstate cooperation (UN-Water 2008). The case studies presented also demonstrate that creating and sustaining TRBOs is a very long process requiring long-term commitment for financial and technical support from the respective governments and other players.

An important message cutting across all the case studies and also articulated in the published literature is that initiatives that enhance transboundary water cooperation invariably contribute to, or even result in, political processes and institutional capacities that open the door to other collective actions, enabling interstate cooperation beyond the river (Wolf 2005). Reduction of tensions among riparian states may also enable joint economic development ventures unrelated to water that would not have been feasible under strained relations, for example, improvements in regional transport infrastructure (Sadoff and Grey 2002). These positive externalities may only be realized if the key actors are capacitated to drive the process forward.

Approaches and imperatives for capacity building in TRBM

In most of the literature and in the case studies presented in this chapter, three key areas of focus in capacity-building initiatives are commonly articulated and promoted in TRBM contexts. The first is human capacity building, whose main product is individuals with the skills to analyze development needs, design and implement strategies and programs, deliver services, and monitor results. The second is organizational capacity building, which produces groups of professionals sharing a common purpose and vision for the river basin, with clear objectives and the internal structures, systems, and resources needed to achieve them (World Bank 2005). The third is institutional capacity building, which focuses mainly on improving the formal "rules of the game" and informal norms that set the boundaries of human action, for example, regulations for water development and use in the river basin (Baser and Morgan 2008).

Since human capacity, organizational processes, and institutional frameworks are closely interrelated and do not change quickly, capacity-building efforts must necessarily be long term and systemic (SIDA 2002; World Bank 2005; BMZ 2006). High priority should also be given to building shared understanding about what works and what does not in terms of improving the enabling environment (OECD-DAC 2006). Contrary to traditional perceptions, this requires a longer-term perspective focused on organizational systems' capacities rather than on individuals or hardware alone (AU/NEPAD 2009). At the individual person level, new water managers should be able to design and facilitate the process of IWRM in transboundary contexts by identifying water-related problems and "hot spots" early on, carefully defining the problem, understanding the interests of all stakeholders, designing appropriate solutions, and facilitating implementation to a satisfactory conclusion (Savenije and Hoekstra 2002).

Creation of TRBOs has often been assumed to be the answer to TRBM challenges in several parts of the world. But this assumption has, in many cases, proved incorrect, and some of the institutions established for this purpose in Africa have remained functionally weak and ineffective (Merrey 2009). Most of the constraints faced relate to the ineffectiveness of leadership and management practices, administrative and delivery mechanisms, information and communication systems, skills and knowledge gaps, and suboptimal allocation and utilization of resources (UNDP 2007). The AU/NEPAD (2009) argues that much of Africa's inability to implement programs with far-reaching impact stems from systemic weaknesses at various levels. Sometimes TRBOs lack the capacity to attain the best possible economic improvements within their respective basins while simultaneously respecting the national sovereignty of riparian countries (Nielsson 1990; Phillips et al. 2008).

There is also a realization that previous efforts and approaches to capacity building have not delivered the desired results and that capacity constraints still remain one of the major obstacles to TRBM in Africa. While some transboundary agreements seem to be effective, some are violated, and others are simply not implemented (UNDP 2007; Vollmer et al. 2009). This implies an urgent need for capacity building if any of the initiatives are to have any positive impact, and efforts by riparian states and donors to build this capacity should continue (Adams 2000; Jacobs 2009). As TRBM capacity building gains momentum, new TRBOs emerge, motivated by sustainable development imperatives and continually "retooling" their business toward a broader mandate of socioeconomic development and ecological sustainability (Hooper and Lloyd 2011).

Programs capacitating TRBO officials should address a mix of knowledge areas and skills that enhance the understanding of basin biophysical, socioeconomic, institutional, and integrative aspects critical for success (Olsen et al. 2006; Douven et al. 2012). FAO (2006) identifies the paucity of accessible international training materials succinctly integrating negotiation skills with international water law as a critical aspect in capacity building for TRBOs. The challenge for the international community is to help develop institutional capacity and a culture of cooperation in advance of costly and time-consuming crises, which threaten livelihoods and regional stability. The academic and professional background of those tasked to lead TRBOs and TRBM processes also matters. The water resources management profession has traditionally been dominated by people coming from engineering and hydrology backgrounds. Typically, such managers will be well versed in water issues but may lack the skills needed to engage in international negotiations, develop legal frameworks for TRBM, and understand international hydropolitics, stakeholder participation, and strategic communications

(Zaag and Savenije 2001; Earle et al. 2010). Targeted capacity building in these domains would be most appropriate for the officials in TRBOs.

a) Role of policies and institutions

In most river basins, a shared regional water and economic development vision can only be crystallized in specific legal frameworks, harmonized policies, and strategic action plans that promote cooperation and also build trust among the riparian states. Therefore, a continuing capacity-building challenge in TRBM contexts is the development and deployment of simple, low-cost but effective institutional frameworks. UN-Water (2008) states that with a few exceptions, such as the Southern Africa Development Community SADC, little attention has been paid to the development of legislative instruments and development of a common vision for sharing water. For example, the NBA and the LCBC are charged with the management of their basin water resources but individual countries still rely on their own setup for managing water resources in their own sections of the basin. According to Sadoff and Grey (2002), international law provides guidance but no clear hierarchy for competing claims on shared waters. Salman and Boisson (2005) state that despite decades of study by the International Law Commission, the international legal community still needs to bring more clarity to both substance and procedure in this area of law. Therefore, capacity-building initiatives targeting policy and legislative development and harmonization across the riparians would add value to sustainable TRBM.

Some scholars argue that the use of existing institutional frameworks established under Regional Economic Communities to achieve the agenda for TRBM and economic integration is strategic, as it is likely to be less costly and more effective than to have custom-built separate structures for each regional water agreement reached (Qaddumi 2008; Bach et al. 2012). By increasing the degree of contact and interaction among member countries, Regional Economic Communities foster trust and mutual understanding, as well as generating practical experiences in problem solving. They can also easily expand the range of potential issues at stake by embedding cooperation over water within a wider framework and increase the possibility of finding a configuration of benefits that is acceptable to all parties (UN-Water 2008; Jacobs 2009).

Defining the interface between shared watercourse institutions and national water management institutions is one of the main institutional challenges for transboundary water resources management in Africa (Savenije and Zaag 2000). Although national water laws often require that international obligations are met, domestic water management institutions are usually not well informed about their role in this process and how it relates to their other obligations resulting from the respective national laws. Wolf (2002) points out that it is crucial for capacity to be built in institutions at all levels to understand the nature of the inter-related governance framework applicable in TRBM and the institutional responsibilities resulting from it in order to ensure the effective implementation of international water agreements. Evidence from various river basins has also demonstrated that constraints to capacity building for TRBM often include institutional arrangements established in an overly top-down manner without recognition of the broader context, and "rigid" allocation of organizational responsibilities reinforcing existing boundaries and undermining interstate collaboration (SLIM project 2004).

b) Power dynamics and neutral dialogue platforms

In the literature, the asymmetry between riparian countries in terms of socioeconomic development, capacity to manage their water resources, and political and military muscle is identified as a challenge to inter-state cooperation, joint management, and protection of transboundary water resources (UN-Water 2008; Hooper and Lloyd 2011). Typically, the more powerful basin states have greater resources to draw on and are thus able to dominate and shape the discourse and control the direction of negotiations (Allan and Mirumachi 2010; Earle et al. 2010; Kim and Glaumann 2012). Even significant gains to cooperation in a river system may not be sufficient motivation for cooperation if the distribution of those gains is, or is perceived as, inequitable (Sadoff and Grey 2002; Zeitoun and Allan 2008). In this chapter, we contend that any perceived unfairness in water allocation and decision-making may lead to disagreements and conflict. At the same time, these differences also open up opportunities for capacity building and more broad-based economic cooperation. They also further open up opportunities for challenging power asymmetries and increasing equity in TRBM.

The absence of institutions specifically mandated to facilitate and coordinate enduring dialogue platforms in African transboundary basins remains a major gap (WWF et al. 2010). In cases where TRBOs already use multi-stakeholder water dialogues, significant progress is notable. For example, the SADC implemented a successful water dialogue approach in 2007 that has been instrumental in serving as a regional, neutral multi-stakeholder platform linking governance levels, water users, and key knowledge generators such as scientists and academics to policy-makers (Bach et al. 2012). The Mekong Water Dialogues, organized by the International Union for Conservation of Nature (IUCN), with a focus on developing and demonstrating participatory processes for improved decision-making in the basin is also another notable example (IUCN et al. 2007). Neutral dialogue platforms present the possibility of enabling all riparians to understand each other's major concerns and their upstream-downstream interdependencies, thereby enabling the crafting of win-win solutions that boost regional integration (SIWI 2009; Chheang 2010). Capacity-building programs that facilitate the development of neutral dialogue platforms are likely to make a positive difference to the TRBM process.

In setting up interstate neutral dialogue platforms, it is also important to note that conflict prevention and resolution are highly political processes in which politicians make decisions on resource use, and political structures of the riparian countries affect TRBM significantly (Katerere et al. 2001; Earle et al. 2010). This suggests that the management of transboundary water is heavily influenced by "hydropolitics." Any attempt to focus solely on building the capacity of TRBOs while sidelining the politicians will be futile (Swatuk 1996; Turton 2002; Jacobs 2009). Therefore, capacity-building initiatives should target the technical experts and politicians simultaneously to develop their international negotiation skills (UNECA 2006).

c) Participation of basin communities and stakeholders

The involvement of key stakeholders in decision-making processes is increasingly recognized by scholars as an important pillar of the institutional framework for TRBM (Ashton 2000; Biswas and Tortajada 2009). Stakeholder participation is fundamental to maximizing agreement, enhancing transparency and decision-making, creating ownership, and facilitating the acceptance and enforcement of decisions and policies. It is also a mechanism for gaining a better or common understanding between the various stakeholders on the nature of a given problem and the desirability of specific outcomes (Wolf 2005). Awareness creation and improved information flow are also likely to be enhanced by the involvement of stakeholders in decision-making processes, particularly at the management level where water allocation decisions are made.

The published literature is indeed awash with specific calls for more local stakeholder and civic participation in TRBM. But operationalizing this "pillar" into a meaningful process that adds value to overall TRBM remains a key challenge, perhaps compounded by the fact that the majority of basin managers and experts tend to be engineers who may not necessarily have the necessary academic training and mind-set for broad-based stakeholder participation. Thus, the actual nature and objectives of stakeholder participation at the transboundary level remain hazy and unclearly defined. An enduring challenge is, therefore, the formulation, incorporation, and implementation of stakeholder participation plans in TRBM. What needs to be determined is how, when, and where to involve these various stakeholders and if there should be common minimum conditions for their participation. Despite the numerous critiques of conventional state-based approaches to TRBM, few among the implementing agencies and their development partners seem to be listening. And it is clear that few experts are thinking in terms of developing these institutions from locally based indigenous foundations (Wolf 2005). Capacity-building initiatives that facilitate and enhance this component will make a big difference to the whole landscape (UN-Water 2008).

d) Protecting ecosystem goods and services

Almost all case studies presented in this chapter reveal that, if no proactive prevention mechanisms are put in place, transboundary river basin economic development projects invariably lead to unsustainable exploitation of the rivers' natural resources. High economic outputs lead to high costs for ecosystems and livelihoods arising from water pollution, overfishing, and highly degraded deltas and coastal zones (Bach et al. 2012). Therefore, solving transboundary environmental problems and protecting ecosystem goods and services is a critical role for most TRBOs. Capacity-building efforts that enhance the TRBOs' environmental management capabilities will go a long way in addressing this challenge (McBeath 2004; Lindemann 2005).

e) Scientific information generation, exchange, and management

Effective TRBM requires that TRBOs individually and jointly generate and share credible data on available water resources with the riparians. In so doing, they enable more effective monitoring of water quantity and quality. Jointly developed plans also have more credibility and effectiveness than plans developed by individual states in isolation (UNECA 2006). In this chapter, we argue that comprehensive capacity-building interventions should seek to improve the competencies and skills of TRBO officials in implementing the broad range of activities constituting integrated water resources management, including application of the transboundary diagnostic analysis framework

(GWP 2012). Other areas of focus include developing the skills to undertake joint basin studies, establish joint databases, quickly exchange information in cases of impending crises such as floods and pollution, establish disaster-preparedness plans, strengthen regional research on topics related to TRBM, and prepare joint basin development plans (BMZ 2006; SIWI 2009).

The need to generate trusted and credible data also magnifies the role of scientists, scientific tools, and volunteer actors. Kranz et al. (2005) state that universities and research institutions now show a very significant production of data and information necessary for TRBM from all kinds of perspectives. They acquire, analyze, and coordinate the primary data necessary for good empirical work, identifying indicators of future water disputes and/or insecurity in regions most at risk. They also train tomorrow's water managers in an integrated fashion. A major challenge for capacity-building efforts then is to ensure that funding for scientific research is sustained over a long period of time until required results are generated and disseminated.

f) Funding

Effective TRBM requires adequate financing. Major costs arise from developing policy and institutional frameworks; developing managerial capacity; creating monitoring, data-gathering, and assessment systems; and creating investment programs that optimize equitable use and protection of the shared water body. While the level of necessary financing varies broadly from one basin to another, sustainable funding repeatedly emerges as a key challenge in most discussions of TRBM (UN-Water 2008). In some cases, TRBOs develop overly ambitious programs that do not focus on priority areas. Administrative, managerial, technical, and financial problems may arise, leading to internal and external pressures that cause poor performance (UNECA 2006). The limited resources available are often overwhelmed. Member states also often fail to sustain the initial political will and commitment by reneging on financial contributions promised. The TRBOs formed become plagued by financial, administrative, and managerial problems (UN 2003). Potential sources of finance for TRBM initiatives range from national budgets and external donor-funded projects to more strategic public-private partnerships. But the funding and investment needs invariably exceed the resources available to riparian countries (UN-Water 2008). The international donor community has contributed important financial and technical support to nearly all TRBOs in the world, and such efforts should continue in Africa.

Conclusion

The exploration of TRBM discourses and capacity-building imperatives done in this chapter reveals that cooperation and not conflict is becoming the norm in TRBM, but basic challenges remain. Capacity-building interventions should target to address challenges and opportunities evident in this landscape to support the broader objectives of regional economic integration. The interventions should seek to change perceptions and capacitate key players in the TRBM sector to understand and apply key tenets of benefit sharing and collective action. The capacity-building objectives set should be realistic rather than be part of a long "wish list" of vague statements from the riparians. This can be achieved through targeted awareness-raising sessions and training programs focusing on proactive conflict prevention.

Trust among riparian countries remains one of the key drivers of effective transboundary water cooperation. Joint basin biophysical analysis, data sharing, planning, and environmental monitoring, which may be enhanced by application of Transboundary Diagnostic Analysis (TDA) and other relevant technical tools, should be promoted in specific capacity-building interventions. However, before the implementation of any intervention, there is the need to analyze in depth the fundamental capacity constraints and challenges that TRBM players face in relation to these key drivers. This enables the design and implementation of more integrated and sustainable capacity-building interventions that address the deep-rooted systemic constraints to functional and institutional performance. Transforming transboundary water institutions is a slow process requiring long-term commitment from governments and their technical cooperation partners. Reliable long-term funding is, therefore, the absolutely necessary ingredient for activities that capacitate people to assume new TRBM responsibilities. If previous efforts and approaches to capacity building have not delivered the desired results and capacity constraints remain one of the major obstacles to African TRBM, then a more systematic approach based on lessons learnt in the past decades will enhance chances of success.

References

Adams, A. (2000). "Social Impacts of an African Dam: Equity and Distributional Issues in the Senegal River Valley." Contributing Paper, Thematic Review I.1, World Commission on Dams, Nairobi: United Nations Environment Programme.

African Union/NEPAD (AU/NEPAD). (2009). *The AU-NEPAD Capacity Development Strategic Framework—Seeing African People as the True Resource*, Midrand: NEPAD Secretariat.

Allan, J.A. and Mirumach, N. (2010). "Why Negotiate? Asymmetric Endowments, Asymmetric Power and the Invisible Nexus of Water, Trade and Power That Brings Apparent Water Security," in A. Earle, A. Agerskog, and J. Öjendal (eds.), *Transboundary Water Management Principles & Practices*, London: Earthscan, pp. 13–26.

AMCOW (2007). *Source Book on Africa's River and Lake Basin Organisations*, Abuja, Nigeria: AMCOW.

Ashton, P.J. (2000). "Southern African Water Conflicts: Are They Inevitable or Preventable?" in H. Solomon and A.R. Turton (eds.), *Water Wars: Enduring Myth or Impending Reality*, *Africa Dialogue Monograph Series No. 2*, Durban: ACCORD, pp. 62–105.

Ashton, P. (2003). "The Search for an Equitable Basis for Water Sharing in the Okavango River Basin," in M. Nakayama (ed.), *International Waters in Southern Africa*, Tokyo: United Nations University Press: 164–188.

Bach, H., Bird, J., Clausen, T.J., Jensen, K.M., Lange, R.B., Taylor, R., Viriyasakultorn, V., and Wolf, A. (2012). *Transboundary River Basin Management: Addressing Water, Energy and Food Security*, Vientiane, Lao PDR: Mekong River Commission.

Bach, H., Clausen, T.J., Trang, D.T., Emerton, L., Facon, T., Hofer, T., et al. (2011). *From Local Watershed Management to Integrated River Basin Management at National and Transboundary Levels*, Vientiane, Lao PDR: IWMI.

Baser, H. and Morgan, P. (2008). "Capacity, Change and Performance Study Report," Discussion Paper No 59B, April 2008, European Centre for Development Policy Management, Maastricht.

Biswas, A.K. and Tortajada, C. (2009). "Water Crisis: Myth or Reality?" *Global Asian Newsletter* (October–December): 1–3.

BMZ (German Technical Cooperation). (2006). "Transboundary Water Cooperation," *BMZ Position Paper*, Special Issue 136 (July), Bonn: 3–16.

Boesen, J. and Ravnborg, H.M. (2004). "From Water 'Wars' to Water 'Riots': Lessons from Transboundary Water Management," Danish Institute for International Studies Working Paper no. 2004/6, Copenhagen.

Burchi, S. and Spreij, M. (2003). *Institutions for International Freshwater Management*, Report for FAO Development Law Series, FAO Legal Office, PC-CP Series no. 3, Rome.

Centre for International Economics (CIE). (2010). *Economic Benefits of Trade Facilitation in the Greater Mekong Subregion*, Canberra: CIE.

Chenje, M. (1996). "Regional Overview: People and Water," in M. Chenje and P. Johnson (eds.), *Water in Southern Africa*, Harare: IUCN, pp. 1–24.

Chheang, V. (2010). "Environmental and Economic Cooperation in the Mekong Region," *Asian European Journal*, 8: 359–368.

Conca, K. (2006). *Governing Water: Contentious Transnational Politics and Global Institution Building*, Cambridge, MA: MIT Press.

Douven, W., Mul, M.L., Lvarez, B.F.A, Son, L.H., Bakker, N., Radosevich, G., and Zaag, P. (2012). *Enhancing Capacities of Riparian Professionals to Address and Resolve Transboundary Issues in International River Basins: Experiences from the Lower Mekong River Basin*, Delft, the Netherlands: UNESCO.

Earle, A., Jagerskog, A., and Ojendal, J. (2010). *Transboundary Water Management*, Stockholm: SIWI.

Falkenmark, M. (1989). "The Massive Water Scarcity Now Threatening Africa: Why Isn't It Being Addressed?" *Ambio*, 18(2): 112–118.

Falkenmark, M., De Fraiture, C., and Viek, M.J. (2009). "Global Change in Four Semi-arid Transnational River Basins: Analysis of Institutional Water Sharing Preparedness," *Natural Resources Forum*, 33: 310–319.

FAO (Food and Agriculture Organization). (2004). *Drought Impact Mitigation and Prevention in the Limpopo River Basin: A Situation Analysis*, FAO Land and Water Discussion Paper 4, Rome.

FAO (Food and Agriculture Organization). (2006). *FAO Training Manual for International Watercourses/River Basins including Law, Negotiation, Conflict Resolution and Simulation Training Exercises*, Rome: FAO.

Fukuda-Parr, S., Carlos, L., and Malik, K. (2003). *Capacity for Development: New Solutions to Old Problems*, London: Earthscan.

Gerlak, A.K. (2007). "Lesson Learning and Transboundary Waters: A Look at the Global Environmental Facility's International Waters Program," *Water Policy*, 9(2007): 55–72.

Global Water Partnership. (2012). *The Handbook for Integrated Water Resources Management in Transboundary Basins of Rivers, Lakes and Aquifers*, Geneva: GWP.

Hall, D. and Bouapao, L. (2011). "Social Impact Monitoring and Vulnerability Assessment: Report of a Regional Pilot Study," *MRC Technical Report, No. 30*, Vientiane, Lao PDR: Mekong River Commission.

Heyns, P. (2002). "The Inter-Basin Transfer of Water between Countries within the Southern African Development Community (SADC): A Developmental Challenge of the Future," in A. Turton and R. Henwood (eds.), *Hydropolitics in the Developing World: A Southern African Perspective*, Pretoria, South Africa: IWMI, pp. 157–176.

Hooper, B.P. and Lloyd, G.M. (2011). *Report on IWRM in Transboundary Basins*, Nairobi: UNEP-DHI Centre for Water and Environment.

International Commission for the Protection of the Danube River (ICPDR). (2009). *The Danube River Basin Management Plan*, Vienna: ICPDR Secretariat.

International Food Policy Research Institute. (2007). *Food Policy Research for Developing Countries: Emerging Issues and Unfinished Business*, Washington, DC: IFPRI.

Isiorho, S.A., Oguntola, J.A., and Olojoba, A. (2000). "Conjunctive Water Use as a Solution to Sustainable Economic Development in Lake Chad Basin, Africa," *Paper presented at the 10th World Water Congress*, Melbourne.

IUCN/TEI/IWMI/M-POWER. (2007). "Exploring Water Futures Together: Mekong Region Waters Dialogue," Report from regional dialogue, Vientiane, Lao PDR, convened by The World Conservation Union, Thailand Environment Institute, International Water Management Institute, and the M-POWER.

Jacobs, I.M. (2009). "Norms and Transboundary Co-Operation in Africa: The Cases of the Orange-Senqui and Nile Rivers," PhD thesis, University of St. Andrews, Ireland.

Katerere, Y. Hill, R., and Moyo, S. (2001). *A Critique of Transboundary Natural Resource Management in Southern Africa*, Harare, Zimbabwe: IUCN—ROSA.

Kim, K. and Glaumann, K. (2012). *Transboundary Water Management: Who Does What, Where?* Stockholm: Swedish Water House.

Klaphake, A. and Scheumann, W. (2006). *Understanding Transboundary Water Cooperation: Evidence from Africa*, Working Paper on Management in Environmental Planning, Technical University of Berlin.

Kranz, N., Interwies, E., and Vidaurre, R. (2005). *Transboundary River Basin Management Regimes: The Orange Basin Case Study*, London: NeWater.

Lindemann, S. (2005). "Explaining Success and Failure in International River Basin Management—Lessons from Southern Africa," *Paper presented at the 6th Open Meeting of the Human Dimensions of Global Environmental Change Research Community*, University of Bonn, Germany, October 9–13, 2005.

MacQuarrie, P.R., Viriyasakultorn, V., and Wolf, A.T. (2008). "Promoting Cooperation in the Mekong Region through Water Conflict Management, Regional Collaboration, and Capacity Building," *GMSARN International Journal*, 2: 175–184.

McBeath, J. (2004). "One Basin at a Time: The Global Environment Facility and Governance of Transboundary Waters," *Global Environmental Politics* (November), 4(4): 108–141.

Merrey, D. (2009). "African Models for Transnational River Basin Organizations in Africa: An Unexplored Dimension," *Water Alternatives*, 2(2): 183–204.

Nielsson, G.P. (1990). "The Parallel National Action Process," in A.J.R. Groom and P. Taylor (eds.), *Frameworks for International Co-Operation*, London: Pinter Publishers, pp. 78–108.

Niger Basin Authority (NBA). (2004). *Shared Vision for Sustainable Development of Niger River Basin, a Strategic and Participatory Approach*, Niamey, Niger: NBA.

Niger Basin Authority/GEF. (2002). *Inversion of the Soil and Water Degradation Tendencies in the Niger River Basin, Trans-Boundary Diagnosis*, Niamey, Niger: NBA.

Odada, E., Oyebande, L., and Oguntola, J.A. (2004). *Lake Chad: Experiences and Lessons Learned Brief*, Washington, DC: World Bank.

OECD-DAC. (2006). "Network on Governance—The Challenge of Capacity Development," in *Working towards Good Practice*, London: OECD: 7.

Olsen, B., Juda, L., Sutinen, J.G., Hennessey, T.M., and Grigalunas, T. (2006). *A Handbook on Governance and Socioeconomics of Large Marine Ecosystems*, Kingston: University of Rhode Island.

Phillips, D.J.H., Allan, J.A., Claassen, M., Granit, J., Jägerskog, A., Kistin, E., Patrick, M., and Turton, A. (2008). *The TWO Analysis: Introducing a Methodology for the Transboundary Waters Opportunity Analysis*, Stockholm: Stockholm International Water Institute.

Qaddumi, H. (2008). *Practical Approaches to Transboundary Water Benefit Sharing*, Overseas Development Institute ODI, Working Paper 292, London.

Rieu-Clarke, A. Moynihan, R., and Bjørn-Oliver Magsig, B. (2012). *UN Watercourses Convention User's Guide*, Dundee, UK: UNESCO.

Saasa, O. (2007). "Enhancing Institutional and Human Capacity for Improved Public Sector Performance," *Paper presented at the UNDP 7th Africa Governance Forum*, Burkina Faso, October 2007.

Sadoff, C.W. and Grey, D. (2002). "Beyond the River: The Benefits of Cooperation on International Rivers," *Water Policy*, 4(2002): 389–403.

Sadoff, C.W. and Grey, D. (2005). "Cooperation on International Rivers: A Continuum for Securing and Sharing Benefits," *Water International*, 30(4): 420–427.

Sadoff, C., Whittington, D., and Grey, D. (2002). *Africa's International Rivers: An Economic Perspective*, Washington, DC: Directions in Development.

Salman, M.A. and Boisson, L. (2005). *Water Resources and International Law*, the Hague: Martinus Nijhoff Publishers.

Savenije, H.H.G. and Van der Zaag, P. (2000). "Conceptual Framework for the Management of Shared River Basins with Special Reference to the SADC and EU," *Water Policy* 2(1–2): 9–45.

Scheumann, W. and Neubert, S. (eds.) (2006). *Transboundary Water Management in Africa: Challenges for Development Cooperation*, Bonn: Ministry for Economic Cooperation and Development (BMZ).

SIWI (Stockholm International Water Institute). (2009). *Getting Transboundary Water Right: Theory and Practice for Effective Cooperation*. Stockholm: Stockholm International Water Institute.

SLIM. (2004). "Stakeholders and Stakeholding in Integrated Catchment Management and Sustainable Use of Water," SLIM Policy Brief No. 2, SLIM, Brussels: European Commission.

Swatuk, L. (1996). "Environmental Issues and Prospects for Southern African Regional Co-Operation," in H. Solomon and J. Cilliers (eds.), *People, Poverty and Peace: Human Security in Southern Africa*, ISS Monograph Series, 4, Midrand, South Africa: Institute for Security Studies, pp. 38–48.

Swedish International Development Agency (SIDA). (2002). *Methods for Capacity Development—A Report for Sida's Project Group: Capacity Development as a Strategic Question*, Working Paper No. 10, Stockholm: SIDA.

Turton, A. (2002). "River Basin Commissions in Southern Africa," *Paper presented at the World Summit on Sustainable Development*, IUCN Environment and Security Day, September 3, Johannesburg.

Turton, A. (2003). "The Hydropolitical Dynamics of Cooperation in Southern Africa: A Strategic Perspective on Institutional Development in International River Basins," in A.R. Turton, P. Ashton, and T.E. Cloete (eds.), *Transboundary Rivers, Sovereignty and Development: Hydropolitical Drivers in the Okavango River Basin*. Pretoria & Geneva: AWIRU & Green Cross International, pp. 83–99.

Turton, A. (2004). "A South African Perspective on a Possible Benefit-Sharing Approach for Transboundary Waters in the SADC Region," *Water Alternatives*, 1(2): 180–200.

Uitto, J.I. and Duda, A.M. (2002). "Management of Transboundary Water Resources: Lessons from International Cooperation for Conflict Prevention," The *Geographical Journal*, 168(4): 365–378.

United Nations Development Programme. (2007). *Building the Capable State in Africa*, Seventh African Governance Forum Report, Ouagadougou, Burkina Faso, October 24–26.

United Nations Development Programme. (2008). *Capacity Development Practice Note*, New York: UNDP.

United Nations Economic Commission for Africa (UNECA). (2000). *Transboundary River/Lake Basin Water Development in Africa. Prospects, Problems and Achievements*, Addis Ababa: UNECA.

United Nations Economic Commission for Africa. (2006). *Africa Water Development Report 2006—Sharing Water for Regional Integration*, The Hague: UNECA.

United Nations (UN). (2003). *International Rivers and Lakes*, New York: UN.

United Nations—Water (UN-Water). (2008). *Transboundary Waters: Sharing Benefits, Sharing Responsibilities*, New York: UN.

United Nations—Water (UN-Water). (2009). *Institutional Capacity Development in Transboundary Water Management*, New York: UN.

UNOPS. (2001). *The Danube River Basin Project*, Geneva: UNOPS.

Vollmer, R., Ardakanian, R., Hare, M., Leentvaar, J., Van der Schaaf, C., and Wirkus, L. (2009). *Institutional Capacity Development in Transboundary Water Management. United Nations World Water Assessment Programme Insights*, Paris: UNESCO.

Wirkus, L. and Böge, V. (2006). *Water Governance in Southern Africa: Cooperation and Conflict Prevention in Transboundary River Basins*, Bonn: BICC.

Wolf, A. (2002). "The Importance of Regional Co-Operation on Water Management for Confidence-Building: Lessons Learned," *Paper presented at the 10th OSCE Economic Forum on Co-operation for the Sustainable Use and the Protection of Water Quality*, Prague, Czech Republic, May 28–31.

Wolf, A. (2005). "Hydropolitical Vulnerability and Resilience: Series Introduction," in A.T. Wolf (ed.), *Hydropolitical Vulnerability and Resilience along International Waters: Africa*, Nairobi: UNEP: 3–18.

World Bank. (2005). *Capacity Building in Africa An OED Evaluation of World Bank Support*, Washington, DC: World Bank.

WWF-DFID. (2010). *International Architecture for Transboundary Water Resources Management: Policy Analysis and Recommendations*, London: Worldwide Fund for Nature and the UK Department for International Development.

Zaag, P. and Savenije, H. (2001). *Conflict Prevention and Cooperation in International Water Resources Course Book*, Delft: UNESCO-IHE.

Zeitoun, M. and Allan, J.A. (2008). "Applying Hegemony and Power Theory to Transboundary Water Analysis," *Water Policy*, 10(Supplement 2): 3–12.

7
The Gas and Oil Sector in Ghana: The Role of Civil Society and the Capacity Needs for Effective Environmental Governance

Cristina D'Alessandro, Kobena T. Hanson, and Francis Owusu

Introduction

The budding commercial exploitation of oil in Ghana has simultane-ously raised expectations and concerns in this African nation, widely considered by donors as a "model country," given its successes in demo-cratic governance and good economic performance despite its internal regional disparities and widespread poverty. Ghana's oil has also con-tributed to the large and growing literature on the multiple challenges of the oil industry in Africa's development efforts. These are indication of the extent of attention that Ghana's oil has been able to attract, both within the country and internationally. However, there is also the widespread fear among many analysts that Ghana's oil manna could turn into a nightmare and paralyze all of Ghana's developmental efforts, particularly those related to its macroeconomic and political stability, potentially threatening internal peace and social cohesion, and stifling the expected economic and social benefits expected from this crucial commodity. Certainly, the lessons learned from other African countries' experiences in the oil sector, the increased international attention, the media's effort in creating awareness and sharing information about this resource, and the roles of civil society and international and local non-state actors have already made a significant difference in Ghana's expe-rience. Furthermore, being well endowed with other natural resources (notably gold, diamond, bauxite, manganese, timber, and rubber) has meant that Ghana has a long experience with the management and exploitation of such resources, especially as pertains to small-scale and artisanal mining. Oil is nevertheless a very peculiar resource. Ghana is making great policy-making efforts, including mining law reforms and changes to investment codes to better cope with this new resource and to ensure that oil revenue is transparently and well spent to benefit the majority of the population. Despite these efforts, the country still faces significant difficulties and challenges in translating its newfound oil riches into development results. Ghana's many civil society organizations (CSOs) have played important roles in this process and, we argue, could potentially play an even bigger role going forward. In fact, CSOs have a long history in Ghana and are important actors in the social and political arena, including the natural resource sector.

This chapter draws on the extant literature to explore Ghana's civil society's role, but more specifically its links to the oil industry. We are especially interested in a particular aspect of natural resource management (NRM) and governance: environmental governance. In the context of this chapter, environmental governance is defined as the "management" of the oil and gas industry with respect to the environment. The concept rests on three key pillars: issues of transparency and economic responsibility, environmental sustainability, and responsible community development (World Bank 2010). Being not only "natural," but also social, environmental governance in the Ghanaian context does include the broader society, as well as specific groups (e.g., fishermen, indigenous communities) that impact and are impacted by the oil and gas industry. Environmental governance is thus linked to social and economic development as much as to environmental concerns.

We argue that even though Ghana's oil is entirely offshore, environmental concerns and risks for biodiversity as well as for communities still arise from its exploitation. Thus, long-term environmental impacts and important risks need to be urgently addressed through a meaningful governance system. Central to the chapter is the following question: What capacities have to be built for CSOs to act effectively and put in place tangible and effective environmental governance for the oil sector in Ghana? To this end, this chapter contends that Ghana has to enhance its environmental governance of the oil resource, and CSOs need to play a central role.

The chapter is structured into seven sections. Following this introduction, the section "Oil in Ghana: A brief background" briefly presents the oil sector and its exploitation in Ghana. The section "Civil society organizations in Ghana" outlines the role of CSOs in Ghana. The section "The Ghanaian civil society in the oil sector: Environmental and social concerns" introduces the environmental and social concerns arising from oil exploitation and the role of CSOs, while the section "Mining, civil society organizations, and environmental governance" defines environmental governance for the oil sector. The section "Capacity needs for enhanced civil society action for environmental governance" analyzes the capacities that need to be built, and the last section concludes the chapter, noting that effective environmental governance for the oil sector in Ghana should be premised on improved capacities of all stakeholders involved in the oil exploitation value chain.

Oil in Ghana: A brief background

The story of oil in Ghana dates back to the 1970s, when commercial quantities of offshore oil reserves were discovered. While production remained negligible for two decades, the hope and efforts by the country to become an important oil producer were never lost. To this end, in 1983, the government established the Ghana National Petroleum Corporation (GNPC) to encourage exploration and possible exploitation of oil in Ghana. The GNPC was also to undertake the development, production, and disposal of oil in the country (Gary 2009; Arthur 2012). As Arthur (2012: 110) further points out, prior to the establishment of the GNPC, the Ministry of Fuel and Power's Petroleum Department was responsible for the procurement of crude oil and petroleum products. GNPC was thus established to support the Government of Ghana's vision of providing adequate and reliable supply of petroleum products and minimizing the country's dependence on crude oil imports. GNPC, by virtue of being the sole custodian of Ghana's onshore and offshore petroleum basins, became the conduit through which any foreign oil company could gain access to petroleum exploration and production rights in Ghana (Gary 2009: 38–39).

In fact, since the 1970s, some oil companies (notably the US-based Amoco, Petro-Canada, the Angolan Sonangol, and the Ghanaian Tema Lube Oil Company) had been involved in the Ghanaian oil sector, because they recognized and understood that there was potential. Similarly, although the GNPC did engage in intensive offshore oil and gas exploration, it made no real headway in terms of actual discovery (Arthur 2012: 111). The country thus had to wait until the mid-2000s to discover commercially viable oil fields and December 2010 to start commercial production from its first oil field, the Jubilee Field, about 60 kilometers offshore in the Gulf of Guinea, in the western part of Ghana not far from Côte d'Ivoire. A consortium of oil companies—Tullow

Oil (UK), Anadarko Petroleum Corporation (USA), and Kosmos Energy (USA)—own this field, while the GNPC holds 13.75 percent equity (Gyasi 2011). The UK-based company Tullow Oil plc, through its subsidiary, Tullow Ghana Limited, is nevertheless the largest foreign actor in the Ghanaian oil sector, owning approximately 35 percent shares of the Jubilee Field—named appropriately so, because the discovery coincided with Ghana's 50th independence anniversary.

Tullow progressively explored new fields, and in March 2009 the Tweneboa discovery was confirmed. Exploration and flow testing have also continued in three close fields as part of a unified project. In November 2012, after some positive discoveries at the TEN project, which is composed of three oil fields—Tweneboa, Enyenra, and Ntomme—Tullow and partners submitted the Plan of Development for this project to the Minister of Energy.[1]

In 2011 Ghana officially became a commercial oil producer, extending the list of African countries that have recently discovered oil. Even if Ghana is only ranked 44th in the world for its proven crude oil reserves, it is ranked 11th in Africa, with approximately 660 million barrels estimated in 2012 (CIA 2013). The production in 2012 averaged 72,000 barrels/day (b/d), but current capacity is put at around 120,000 b/d. This figure can be greatly expanded once more oil blocks enter production. Significant quantities of natural gas have also been found with the oil reserves: about 800 billion cubic feet of gas in the Jubilee Field (Gary 2009).

The Jubilee Field, with its estimated 3 billion barrels of recoverable oil (Arthur 2012), straddles two oil blocks: the West Cape Three Points block led by Kosmos Energy, and the neighboring Deepwater Tano block led by Tullow Oil, which holds approximately 35 percent ownership (Gary 2009).

Oil production in Ghana has resulted in a fast-increasing oil export, yielding US$3,120 million in revenue in 2012 (The Economist Intelligence Unit 2013). Ghana's annual gross domestic product (GDP) growth also rose from 4 percent in 2009 to 14.4 percent in 2011; a rise directly linked to oil and confirming that oil exploitation can significantly contribute to setting in place positive economic trends.

The World Bank (2009) estimates the potential government revenue related to oil at approximately US$1 billion per year on average. This valuation, based on a price assumption of US$75 per barrel and the fiscal regime in place at that time, is probably below the actual contribution of oil to government revenue, but nonetheless considered significant. In fact, government revenue topped US$3.7 billion (excluding grants) in 2008 (World Bank 2009). The International Monetary Fund (IMF) has also predicted that government revenue from oil and gas could reach a cumulative US$20 billion over the production period of 2012–2030 for the Jubilee Field alone. The government's take was estimated at around US$1.3 billion in 2013, more than cocoa and gold earnings combined (Gary 2009).

The aforementioned notwithstanding, critics are quick to point out that the deal between Kosmos and Ghana was too favorable to Kosmos, in essence shortchanging Ghana on potential revenue (Arthur 2012). Yet, others defend the deal, noting that prior to Kosmos' gamble on a deepwater project with no firm guarantee of returns, no major oil company was willing to do so (Kapela 2009; Arthur 2012).

The effects of the oil value chain on the economy and on social activities however go far beyond the aforementioned direct macroeconomic impacts. Even if oil upstream activities are expected to remain limited in the short and medium term, because of the offshore nature of oil exploration and production in Ghana, downstream activities can drive industrial development (not only refineries, but also local content, and local sourcing of goods and services), helping to create employment opportunities and to diversify the economy, as well as to build and improve infrastructure. If well spent, oil revenue can also greatly contribute to building local capacities through education, training, and research, reducing poverty and improving the overall quality of life in the country. As Glyfason (2001: 851) points out, the importance of investing in human capital and capacity development cannot be underestimated in the effort to ensure the resource boom does not result into a resource curse (see also ACBF 2013). In fact, oil revenue stream can contribute not only to fiscal consolidation, but also to finance the Growth and Poverty Reduction Strategy, helping the country to move to a middle-income status. Nevertheless, experiences in the oil sector in other African countries show that this goal is not so easy to reach: external factors, multiple stakeholders and interests involved in the process, the need for equipment and technical skills not locally available, and corruption and lack of transparency can undermine the process if the requisite leadership, institutions, and capacities are not put in place (ACBF 2013).

The aforementioned notwithstanding, concerns still do exist in Ghana regarding the development of the oil industry and the management of oil revenue by the government (Gary 2009; Kapela 2009). Despite the great expectations that are always associated with a strategic industrial activity like oil, which is capable of transforming the national economy, some important developmental objectives such as employment creation, income distribution, social development, and environmental protection, among others, must be deliberately ensured. In fact, neglecting these issues can endanger national peace and harmony, create political instability, and lead to social divisions and conflict. The Nigerian case is a reminder that the "resource curse" and "Dutch disease" must be avoided, that social and spatial redistribution of oil wealth should be a high priority, and that diversifying the economy to avoid overreliance on oil should be the guiding principle of policy-makers (Arthur 2012; ACBF 2013).

Besides the economic consequences of oil exploitation, the Nigerian case, even if different from that of Ghana, also reminds us that environmental and social concerns related to oil activities can be heavy. The disruptive impacts of oil, such as those linked to economic effects of deepening inequalities, may endanger the political system and the governance structure of the state, sow the seeds of local and national conflicts, and threaten peace and stability in the country. These concerns will be explored in the Ghanaian context in the following sections, with emphasis on the role of civil society in the oil sector.

Civil society organizations in Ghana

CSOs are organizations that act in the public arena and represent the interests and values of social groups that are vulnerable, or not powerful enough, to make their voice count in social contexts. CSOs are therefore necessarily formal groups, and diverse for size and scope; however, their existence has to be officially recognized in some way to make their action legitimate. Civil society groups are at the heart of democratic processes, because they allow a large range of stakeholders to participate in policy-making and ensure the implementation of policies, because of the involvement of the social groups that are affected by these policies. Besides technical policy-making, CSOs participate in development processes, in community mobilization and organization in different situations, in research and documentation, in advocacy and popular campaigns, and in improved reporting. In sum, they are both upstream and downstream actors in democratic life.

CSOs have been reported to be in existence in Ghana since 1781 (when the colonial territory was named Gold Coast). The community-based organizations documented during this era included the Fante Confederacy and the Aborigines' Rights Protection Society (Darkwa et al. 2006). After independence, several CSOs were created, but were primarily placed under the ruling party and state supervision and were therefore used as indirect tools of power. There is then a long history and tradition of CSOs in the country, but the contemporary part of it can be considered as starting in the 1980s.

The structural adjustment program (SAP) era, which began in Ghana in the early 1980s, also saw significant increased interest in CSOs, as important partners in development programs: the range of CSO actors continued to widen. In 1987, the Civil Society Coordinating Council (CivisoC) was established with the purpose of representing civil society in the SAP process (Darkwa et al. 2006: 24). The presence of CSOs in the political context has been particularly strong since the 1990s, mainly due to the changing political landscape in the context of the Fourth Republic. This was the beginning of a new era for Ghana's civil society and a real turning point—a large number of CSOs have proliferated in the country since then.

Despite the strong and dynamic tradition of associational life in Ghana, participative mechanisms are in fact, as elsewhere, never completely democratic; some groups and interests are more able than others to represent themselves. For instance, middle-class-run CSOs are more likely to be included in participatory processes, while rural illiterate remote groups are typically disadvantaged during consultations that are often done in English language and take place in the capital city of Accra. Because of these limitations, rooted in the peculiar Ghanaian history, public participation, in Ghana as elsewhere, is never total and "fair," because societies are always uneven groups. It is important to note here that there is a recent but relatively important history of civil society participation in the political life of the country and that Article 276 of the 1992 constitution also guarantees this participation. This article shows that there is awareness, a strong determination, and also political mechanisms defending public participation in political decision-making (Kuyini Mohammed 2013). In fact, apart from financial constraints limiting CSOs' effective functioning in Ghana, decreasing also their capacity to practically play a central role in influencing public policy, people's grievances are likely to find voice among CSOs in Ghana: this explains the high public trust for civil society in the country.

Despite structural limitations common to most African countries, civil society in Ghana has a recognized capacity to challenge state processes and to participate in the political life, especially when decisions are related to NRM and sensitive social topics. Nevertheless, their limited institutional capacities, together with limited financial resources, push them to rely uniquely on activism for their action. Activism can certainly be effective for achieving objectives (Arthur 2012), but cannot be the only strategy of action for civil society. In the oil sector, widespread new technologies, such as the Internet and cell phones, help to share knowledge, to gather consensus, and to widen capacity of action and activism. Oil, a recent and strong economic activity in Ghana, is then easily mobilizing voices and action, but action has to go beyond activism.

The Ghanaian civil society in the oil sector: Environmental and social concerns

In 2007, Ghana discovered commercially viable quantities of oil and gas, with estimated revenue of oil reserves of approximately US$1 billion per year (a figure almost similar to how much Ghana received in development assistance annually). To date, the oil industry in Ghana has been primarily offshore. Even if it is generally acknowledged that offshore drilling and production activities have more limited social and environmental consequences than onshore oil industries, it must be noted that offshore fields must necessarily have an onshore base, with mid- and downstream oil activities. Thus, the socioeconomic and environmental impacts of these activities have to be taken into account, as part of the overall impact of the oil industry in Ghana.

Without doubt, when managed well, the extractive sector, such as oil and gas, can bolster development efforts and advance economic growth and transformation (ACBF 2013; ECA/AU 2013). Doing so, however, calls for a dynamic oversight regime with bodies such as parliaments, media, and civil society playing a critical role in ensuring good governance of the sector.

Civil society can play a critical role in fostering sound management of the sector. However, in Ghana, at the time of oil discovery in 2007, very few CSOs had the requisite capabilities and specific experiences or expertise in this regard (Zandvliet 2013). In order to advance their capacity to engage on oil and gas issues across the entire value chain, CSOs in Ghana established the Civil Society Platform on Oil and Gas (CSPOG) (Zandvliet 2013: 2), uniting 135 CSOs. The CSPOG was created in 2007, with the support of many international actors, including Oxfam America, the European Union, and the World Bank. Ghana is thus following the example of other African oil producer countries, where similar platforms have been created: the Civil Society Coalition for Oil in Uganda, and the Kenya Civil Society Coalition on Oil and Gas. The goal of the CSPOG is to share information about the oil sector among CSOs and to engage more effectively in advocacy and coordinate their activities.

The CSPOG has already been successful in advocating the adoption of legislation in line with best international practices, and in lobbying the government and the parliament to draft and approve two bills, the Petroleum Exploration and Production Bill and the Petroleum Revenue Bill, both passed by the parliament in 2010. It is now working to influence the management and allocation of oil revenues, through the Public Interest and Accountability Commission (Zandvliet 2013: 3). The commission also participates in debates on spending of petroleum revenues in line with development priorities.

Despite the recent progress made since the CSPOG was created, in terms of increased coherence, coordination, and effectiveness of civil society action, important limitations still exist. Ghanaian civil society in the oil sector is still constrained by weak overall technical capacities and insufficient funding. CSOs' action has to be especially reinforced to increase their capacities to engage local communities in monitoring and advocacy, when their interests are attacked. CSOs also have an important monitoring role of government activities in the sector that may involve "lobbying of governmental institutions and agencies in their engagements with the state" (Idemudia 2009: 16, cited in Arthur 2012: 117). Informing and educating the different stakeholders about oil issues is another crucial CSO activity that should be enhanced in Ghana; in fact, this contributes to empower the media and build its capacities to collaborate and support CSOs' action. In the end, through all these actions, civil society also creates the preconditions to promote dialogue between the different stakeholders involved in oil exploitation.

With regard to the issues related to oil as in other domains, CSOs are the interface stakeholders, intermediaries between the government and local communities. They disseminate information both ways (top-down and bottom-up), contributing to inform policy-making with local demands and concerns and helping communities to check and eventually react if state action is not satisfactory, according to their criteria. If appropriate technical, scientific, and vocational skills are then necessary to perform effectively this complex action, international connections and support are also crucial to this extent.

The CSPOG is in fact concerned by its poor ability to tackle environmental concerns (CSPOG 2011). The environmental risks and impacts of oil being so crucial for Ghana in the short and even longer term and not only for the Western Region, enhancing CSOs' capacities in this domain is a priority to ensure environmental and social protection in the oil industry. Environmental concerns about the commercial oil exploitation were raised in 2009 by some members of the above-mentioned platform, including the Integrated Social Development Center (ISODEC), which also serves as the Ghanaian branch of Publish What You Pay, when the Jubilee fields project was developed and when the environmental impact assessment (EIA) of the oil companies themselves showed that the situation was problematic even before the oil production had started. Minor oil spills have also already been reported at the Jubilee Field,[2] and of course the fear that major spills could happen in Ghana exists. It is not even sure that the country would be prepared and equipped to face major oil spills and to avoid major environmental disasters.

In 2011, the platform issued a "Readiness Report Card," which, among others, evaluated the performance of the Government of Ghana in managing the challenges of the emerging oil sector and drew attention to issues requiring immediate action (Zandvliet 2013). The overall scores obtained and reported in the report are generally fair, but even poor for social and environmental concerns and for addressing local concerns. The report underlines the lack of technical capacities, undermining CSOs' action in the oil sector, especially in environmental monitoring: this lack of monitoring skills will be more and more problematic with Ghana moving further in the development of legal frameworks, because a stronger monitoring regime will be required (CSPOG 2011). Environmental and social monitoring is especially crucial to avoid social conflicts, related to impacts of oil exploitation on fishing and agricultural activities, already arising in the Ahanta West District for instance (Agyei et al. 2012a).

According to Kjeldsen (2010), some marine and coastal environments already experienced damages and changes in their flora and fauna as a result of oil activities. Some of these impacts are potentially long-lasting, causing biodiversity losses and human health concerns through irrigation and drinking water. Dangers linked to the use of oil tankers that do not meet international security standards have also been identified and denunciated by nongovernmental organizations (NGOs) and the International Maritime Organization (Kjeldsen 2010). The environmental impact assessment (EIA) developed by oil companies to evaluate the environmental consequences of the Jubilee Field has been criticized, and several weaknesses in the assessment have been identified in a study commissioned by Oxfam America in 2009. According to this study, the EIA is not complete, based on secondhand or not-updated data, and inadequate for the environmental damages caused to the fauna. Plus the EIA does not assess oil spills (Gary 2009).

A recent study on the environmental impact of oil and gas activities in Ghana (Kumar et al. 2013: 347) concluded that Ghana's "marine water quality was affected. The discharge of polluted fluids with the metals composition affects human socio-economical and sea resources. Also the distribution of the iron content affects ground water storage in the specific areas of Ghana. Oil and grease quantity affects mainly organisms in ocean." The above findings concur with the findings of Sakyi et al. (2012: 70), who, in a similar study, pointed out that the "exploration, development and production of oil and gas in the Jubilee Oil Field could be associated with ecological degradation…," and thus called on the Ghana government to take active steps to "develop petroleum industry specific environmental protection guidelines and appropriate regulatory infrastructure including monitoring equipment, compliance enforcement networks and also a deterrent sanction regime."

Very serious risks to the livelihoods of local fishermen have to be considered as crucial problems, too. In fact, depleting fish stocks, partly due to oil exploration and exploitation activities, contribute to explain why local fishermen along the coast struggle to sustain their livelihoods. In a recent study by Agyei et al. (2012b: 186–188), feedback from 200 local fishermen from ten fishing communities of the Ahanta West District in the Western Region of Ghana suggests that "about 25% [of surveyed fishermen] alleged the 500 m security radius zone for the oil rig and 1000 m for the FPSO had caused reduction of fish catch." Beside the quantities of fish, the oil equipment present in the sea has been reported to practically influence the fishing activities. Indeed, many

> fishermen [operating in the locality] complain that the fish at night are attracted by the lights from the oil platforms and hence move towards areas where the fishermen are prohibited from going. There have already been several cases of fishermen coming too close to oil platforms which have led to clashes between them and patrolling vessels. In some cases fishermen have had their catch or nets confiscated, which constitutes a huge financial loss to the individual fisherman.
>
> (Kjeldsen 2010: 22–23; see also Agyei et al. 2012b)

These incidents could have potentially devastating consequences as increased poverty among vulnerable groups such as fishermen could lead to overfishing or massive exploitation of marginal fishing waters (Kjeldsen 2010).

Mining, civil society organizations, and environmental governance

Mining activities generally have heavy environmental impacts, but the scale of these impacts often depends on the type of activity and the way it is undertaken (for instance, commercial or artisanal mining, offshore or onshore). Besides these differences, environmental damages, such as water and air pollution, noise and various disturbances, biodiversity destruction, health concerns and risks, and consequent impacts on livelihoods and income need to be addressed to ensure that the sector is developed in a sustainable manner. In fact, even if environmental consequences cannot be avoided, they have to be at least known and controlled, and legal instruments put in place to protect both the biodiversity and the local populations living and earning their livelihoods in the area. The global endorsement of this principle nevertheless carries problems about its applicability to extractive activities: the numerous and diverse economic interests linked to these activities and the relative short time of their exploitation may be certainly quoted as a reason. For instance, since minerals are nonrenewable resources, full adherence to this principle can limit the growth of the industry and hence its sustainability. This concern has been already raised in the past in Ghana by stakeholders in the mining industry, in connection with attempts to mitigate the negative effects of small-scale gold and diamond mining in the country. Ghana has in fact established a legal framework in this domain since 1989 (the Small-Scale Gold Mining Law 1989 (PNDCL 218)) (Amankwah and Amin-Sackey 2003). The same kind of environmental concerns arise also from commercial large-scale mining activities, such as for oil, even in terms of offshore fields, which generally engender minor direct environmental impacts, compared to inland oil fields. It is in light of the aforementioned that a functional framework for environmental governance that addresses these potential concerns arising from Ghana's emerging oil industry is needed.

In fact, during the 2012 Society of Petroleum Engineers and Australian Petroleum Production & Exploration Association (SPE/APPEA) International Conference "Health, Safety and Environment in Oil and Gas Exploration and Production" in Perth, Australia, Ghana's Minister of Environment Science and Technology noted that "the country's regulatory environmental governance model is evolving toward a consistent global benchmark" (Modern Ghana News 2012). The minister further stressed the importance of integrating ideas that would ensure increased shared understanding on regulatory challenges as being vital to the successful industry and regulation achievements and policy innovations.

In many countries, a dedicated institution is often charged with the management of environmental and social impacts of the oil and gas industry—either the ministry of environment, or a similar institution (World Bank 2010: 9). That said, the effectiveness of a regulatory framework will be "compromised by the lack of a sufficiently organized administrative structure that enables efficient regulatory compliance and enforcement" (World Bank 2010: 10). Another potentially compromising factor of regulatory effectiveness is the lack of human and financial resources required for effective environmental governance (World Bank 2010).

In Ghana, the government has put in place an environmental legislation framework, which has progressively enhanced environmental governance over the years. The Environmental Protection Act of 1994 (Act 490) and the subsequent creation of the Environmental Protection Agency (EPA) are central to this process. Even if there are regulatory gaps and an absence of specific regulations for the oil sector in Ghana, the government shows a will to be part of the environmental governance process (Darkwa et al. 2006). CSOs, such as Oxfam and Friends of the Nation (FoN), are also actively engaged in the national policy dialogue on oil and gas.

As noted earlier, environmental governance is part of the new NRM model. This model is in essence the sustainable management of good natural resource governance, and particularly concerned with improved and informed leadership and strengthened state-civil society-private sector partnerships and cooperation (ACBF 2013). NRM is, then, driven not only by economic criteria, measured through revenues, taxes, royalties, and outcomes generated by the resource, even though these aspects are strategic, but also by environmental and social criteria participating in the overall evaluation of the sector.

Environmental governance seeks to bring on board the different stakeholders involved at various levels of the resource exploitation, especially those that are more vulnerable and not powerful enough to defend their interests in NRM. As Lockwood et al. (2010: 8–9) argue, inclusiveness is a key principle of environmental governance:

> Governance is regarded as inclusive when all those with a stake in governance processes can engage [with each other] on a basis equal to that provided to all other stakeholders. As solutions to NRM problems

often demand substantial changes in practices, their implementation requires [the] participation of as many of the affected actors as possible [more so] because no single actor has the resources to generate solutions to wicked problems.

To set in place such inclusive process, it has to be recognized that society is not an undifferentiated body, but is rather made by the differentiated social actors related to the resource in question (oil and gas in the context of this chapter). Accordingly, the environment is a set of goods and services, some of which in precise contexts become endowments and the entitlements once exploited.

Leach et al. (1999) proposed a framework that underlines the role of institutions in the process: at every scale (from macro to micro), different types of formal and informal institutions and at different levels of the process participate in effective environmental governance. These capabilities of all the social actors involved in the transformative process of endowments are linked to the institutions (being then institutional capacities) and are crucial for the new NRM, aware of environmental concerns. As they further contend:

> An understanding of social difference, and the diverse institutions which support different people's endowments, entitlements and environmental management, points toward possibilities for more strategic specificity in interventions. If certain institutions can be identified as supporting the interests of certain social actors, or as contribution to "desired" courses of ecological change, then they can be targeted by policy in strategies of institution-building or support.
>
> (Leach et al. 1999: 240–241)

CSOs are an integral part of *a corporate social responsibility*: this is a corporate social reporting and multi-stakeholder dialogue between companies, CSOs, and the government (Williams 2004). As Williams (2004: 461) argues, "citizen's demands for corporate social responsibility... are changing the norms of appropriate industry action with respect to important questions such as environmental protection, security arrangements for pipelines and plants, and financial arrangements with host countries." She further adds that corporate social responsibility is in fact part of a "new governance" pattern, in the relations between civil society and the business community.

Environmental governance is then a complex process, in which multiple stakeholders are involved. Oil companies, Tullow for instance in

Ghana, are increasingly involved (as a result of international pressure) in environmental monitoring and assessment (AfDB 2009): they of course try to defend their interests to exploit the oil, minimizing in their statements the impacts of their activities on the environment (CSPOG 2011). Tullow contends that the Jubilee oil field is one of the best fields in Africa in over a decade. Through the environmental and social impact assessment (ESIA), made in 2009, Tullow noted that actions are taken to mitigate its impact on the environment, as well as a series of management plans aimed at ensuring compliance with the environmental and social requirements of operations.[3]

The above discussion points out the crucial role of CSOs and related institutions, and the need to enhance their capacities in acting as stakeholders, having a full part to play, defending the rights of the people that they represent, supporting their claims, and acting in the right institutional contexts. The efficiency and capabilities of institutions at every level are crucial to avoid ambiguities in the determination of people's rights on endowments, and later entitlements, which can create conflicts related to resources (see Fred-Mensah (1999) on land conflict in rural Ghana). Similar arguments can be made for oil management: capable CSOs can reduce these ambiguities and consequently lower or avoid conflicts.

Capacity needs for enhanced civil society action for environmental governance

The CSPOG and the broader civil society involvement in the Ghanaian oil and gas sector cannot play entirely their role for enhancing environmental governance, because they are affected by a number of capacity gaps, the main ones being those listed below:

1) *Capacities to inform communities and manage expectations.* To avoid or limit widespread disappointment and frustrations in local communities, which can degenerate into conflicts, communities have to be provided with relevant in-depth information in order to create realistic expectations (Yeboah et al. 2012). This has to be done through the media, but also through meetings with chiefs and local leaders: discussions and exchanges have to progressively create consensus; after that, communities know the situation and the possibilities and can then negotiate with other stakeholders. This can be the case, for example, for pollution or noise related to oil exploitation: communities have to know their rights and what to do to defend them. They have to be aware of dangers, of possible

consequences engendered by technologies and materials used, and of eventual measures to be taken. NGOs, especially large international organizations, may be of great help, sharing their long-term experience to this extent.

2) *Monitoring capacities.* The capacities of CSOs in general and of the CSPOG in particular to undertake evidence-based environmental and social monitoring in the six coastal districts of the Western Region is really limited. A report by ISODEC and Oxfam America suggests that a GIS map including all the key areas of biodiversity and the fish breeding grounds should be set in place. This is nowadays a standard process to build a model studying the eventual impacts of spills on the marine environment (Gary 2009). Without any data, sample analysis, baseline studies, or other research to prove that oil extraction has negative impacts on the environment and on social activities; CSOs are disempowered in their negotiations with other stakeholders. The ability to collect proofs, to involve communities, and to stimulate community monitoring is a crucial capacity to develop for Ghanaian CSOs and for the CSPOG. It is specifically important to monitor the impacts of oil activities on vulnerable social groups (women, youth, fishermen, etc.), and the CSPOG has to share any monitoring results with oil companies and with local governments on a regular basis and with the parliament for national issues, especially when related to the need for new policies or for changes in existing policies. The collaboration with the EPA, the leading public institution for environmental protection in Ghana, is strategic to enhance these capacities.

3) *Capacities to propose mitigation measures.* Precise monitoring implies another capacity to call for clear and adapted mitigation measures in a timely way, which means a capacity to react promptly and to know the measures proposing the right and possible procedure. This means that knowledge about other similar previous or contemporary situations is shared by the CSPOG with international NGOs and other oil producer countries around the globe, but also that the legislation is adapted to call oil companies or other stakeholders for mitigation measures. Knowledge sharing is for sure easier among countries that have already analogous platforms on oil and gas, and it can be interesting to build on success stories or mistakes of other African countries.

4) *Negotiation capacities.* Dealing with multiple stakeholders, including important private sector actors like international oil companies, with a large experience to handle conflicts and to find legal ways to advantage themselves requires important negotiation

skills that the CSPOG needs to possess. These skills can be gained through international NGOs or also through ad hoc training. International institutions, such as African Development Bank (AfDB) and the African Capacity Building Foundation (ACBF), but also the World Bank, can offer expertise and advice on how to gain these skills in a rapid and effective way. Other organizations and institutions that have expertise in this domain, as the International Council on Mining and Metals (ICMM) or the International Finance Corporation's Oil, Gas and Mining Sustainable Community Development Fund (IFC CommDev), can help improve negotiation capacities.

5) *Capacities to influence policy-making.* Ghana's laws being not adequate to address sea pollution and oil spillages, the country needs to protect itself with a sea pollution law adapted to its needs. CSOs need to support the EPA, which does not have any experience in this domain. More in general, environmental laws and regulations need to be reviewed and expanded to be able to address the oil sector. Policies at the national level should include early warning systems, and special attention has to be paid to harmonize regional policies to national laws. To this extent, the CSPOG has to play a central role in contributing to this process, and pushing for these measures to be undertaken and implemented quickly: extensive reviews of other national legislations could help to do this. International legal expertise, from international organizations for instance, may be called on board.

6) *Capacities for regional cooperation and information sharing for risk management and prevention.* International initiatives of geospatial data infrastructure allowing to share data among numerous countries and to set in place common mechanisms of alert exist around the world. The Global Monitoring for Environment and Security GMES and the Infrastructure for Spatial Information in the European Community (INSPIRE) in Europe are regional examples, while the Global Disaster Alert and Coordinated System (GDACS), established in 2004 by the UN, is a global cooperation framework. All these systems allow providing alerts through in situ captors and impact estimations after disasters. By enhancing the cooperation of international actors and neighboring countries concerned by such possible accidents, the response is certainly more efficient. They can also provide models for elaborated predictions and simulations: they are then useful before, during, and after a disaster (Figure 7.1).

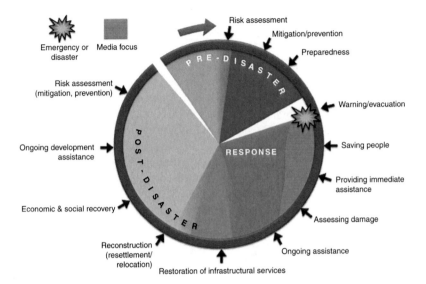

Figure 7.1   Risk cycle diagram
Source: Adapted after Mensah (2010).

The CSPOG should encourage such initiatives in Ghana, using international connections to establish such kind of regional cooperation in the Gulf of Guinea, especially knowing that all Ghana's coastal neighboring countries (Benin, Togo, Côte d'Ivoire, Liberia, Nigeria) are oil producers. Such collaboration would be crucial to protect not only the environment, but also the populations, guaranteeing their source of income. It would also decrease potential conflicts among neighbors, enhancing collaboration, while conflicts between Ghana and Côte d'Ivoire are possible for oil reserve located right at the borders between the two countries, creating tensions for the rights on their exploitation.

Again, the CSOs operating the oil and gas field can seek to better influence the environmental policy and legislation through effective dissemination to decision-makers and citizens. Indeed, given that capacity development for the environment (CDE) is best viewed as an endogenous process of change that operates at multiple levels—enabling environment, organization, and the individual—capacity development "is not an end in itself, and as such environmental goals should be the basis for determining capacity requirements, which in turn should be the basis for defining capacity development priorities" (Matheson and Giroux 2010: 3; see also ACBF 2013).

7) ***Partnership and funding capacities***. The CSPOG and Ghanaian CSOs in the oil and gas sector in general are affected by limited funding available for their action. These reduced financial resources decrease also their ability to build their own capacities. They need to possess a clear and robust funding strategy in the long term, ensuring them at least a minimal level of financial resources, a certain security and stability, and a longer-term action. Nowadays, this can be done through partnerships with funding institutions and through other strategies to earn financial benefits. This is a necessary starting point, because without any financial security the actions undertaken in Ghana cannot have a follow-up, so they can also not be taken seriously by other stakeholders involved in the sector.

Conclusion

The individual and collective capacities of the myriad actors and stakeholders involved in high-value natural resource extraction, processing, marketing, and management of revenues are of fundamental importance in ensuring that resources end up benefiting the broader society (ACBF 2013). Such capacity, however, ideally needs to be balanced between stakeholders. There is a good deal of emerging evidence that "capacity imbalance"—whereby one set of stakeholders enjoys significant capacity while other sets experience lower and in some case much lower capacity—can result in tensions due to a lack of effective checks and balances between sets of stakeholders (ACBF 2013). This chapter, in examining the importance of CSOs in ensuring natural resource governance in Ghana's oil and gas sector, also acknowledges that a great deal of capacity is also required on the part of the international investor, who in many cases is unable or unwilling to "read" local sociopolitical, ethno-cultural, and economic environments so as to be able to innovate and derive arrangements that work and are mutually beneficial—as evidenced by the tensions over territory and environmental issues bubbling up between oil giants and local communities in Ghana (see Kjeldsen 2010; Agyei et al. 2012a).

To this end, Ghana needs to further enhance its efforts at effective governance for oil and gas. Despite the progress made in a short time in the oil and gas management in the country, a number of uncertainties and challenges persist, especially linked to social and environmental concerns. Communities' expectations from the sector, international attention on environmental concerns and regulations in Ghana, and oil companies with their interests and their capacities to negotiate favorable conditions for their work play important roles in the Ghana's oil and gas industry (Agyei et al. 2012b; Arthur 2012; Sakyi et al. 2012).

In this specific context, CSOs have a critical role to play and can make a huge difference in the future of the oil industry in Ghana, by ensuring that social and environmental concerns are at the heart of this value chain. This is possible due to the country's favorable political and economic context, the relatively late discovery of oil (benefiting from previous and contemporary experiences, on the continent and beyond), and also the long tradition and action of CSOs in the country.

For this to happen in Ghana in the short term, CSOs and the CSPOG need to strengthen their capacities in many different ways and directions (some of them have been emphasized above). However, they cannot effect this change by themselves; rather, they need support, collaboration, and partnerships with international institutions and organizations. If knowledge sharing is crucial for enhancing their capacities, the first starting point, the premise to everything, is the capacity to ensure funding in the long term for their action and for building further capacities.

Notes

1. See www.tullowoil.com/ghana/
2. See http://www.wakeupghana.com/2010/01/07/oil-spills-at-jubilee-oil-field .html
3. See http://files.thegroup.net/library/tullow/crreport2009/pdfs/tullowcr09_ jubilee.pdf

References

ACBF (African Capacity Building Foundation). (2013). *Africa Capacity Indicators Report 2013: Capacity Development for Natural Resource Management*, Harare, Zimbabwe: ACBF.

AfDB (African Development Bank). (2009). Ghana Jubilee Field Project. Summary of the Environmental Impact Assessment, October.

Agyei, G., Gordon, J., Erasmus, A., and Yakubu, I. (2012a). "Oil Industry Activities in Ghana: Community Perceptions and Sustainable Solutions," *Research Journal of Environmental and Earth Sciences*, 4(5): 583–596.

Agyei, G., Gordon, K.J., and Adei, I. (2012b). "Offshore Oil Industry Activities and Fishing in Ghana: Community Perceptions and Sustainable Solutions," *Current Research Journal of Social Sciences*, 4(3): 182–189.

Amankwah, R.K. and Amin-Sackey, C. (2003). "Strategies for Sustainable Development of the Small-Scale Gold and Diamond Industry of Ghana," *Resources Policy*, 29: 131–138.

Arthur, P. (2012). "Avoiding the Resource Curse in Ghana: Assessing the Options," in M.A. Schnurr and L.A. Swatuk (eds.), *Natural Resources and Social Conflict. Towards Critical Environmental Security*, Palgrave: Basingstoke, pp. 108–127.

CIA. (2013). *The World Factbook 2013–2014*, Washington, DC: Central Intelligence Agency.

CSPOG (Civil Society Platform on Oil and Gas). (2011). *Ghana's Oil Boom. A Readiness Report Card*, Available at: http://www.oxfamamerica.org/static/oa3/files/ghana-oil-readiness-report-card.pdf. Accessed August 10, 2013.

Darkwa, A., Amponsah, N., and Gyampoh, E. (2006). *Civil Society in a Changing Ghana. An Assessment of the Current State of Civil Society in Ghana*, Accra: CIVICUS/WB.

ECA/AU (Economic Commission on Africa/African Union Commission). (2013). *Economic Report on Africa 2013: Making the Most of Africa's Commodities: Industrializing for Growth, Jobs and Economic Transformation*. Addis Ababa, Ethiopia: ECA/AU.

Fred-Mensah, B. (1999). "Capturing Ambiguities: Communal Conflict Management Alternative in Ghana," *World Development*, 27(6): 951–965.

Gary, I. (2009). *Ghana's Big Test: Oil's Challenges to Democratic Development*, Oxfam America and the Integrated Social Development Centre, Available at: http://www.oxfamamerica.org/files/ghanas-big-test.pdf. Accessed August 10, 2013.

Glyfason, T. (2001). "Natural Resources, Education and Socieconomic Development," *European Economic Review*, 45(4–6): 847–859.

Gyasi, S. Jr. (2011). "Making Oil a Blessing in Ghana," *New African*, February (503): 47–49.

Kapela, J. (2009). "Ghana's new oil: Cause for jubilation or prelude to the resource curse," Unpublished MA thesis, Duke University, USA.

Kjeldsen, U.B. (ed.) (2010). *A Brief Guide for Civil Society to Oil and Gas in Ghana*, Accra: CSPOG and IBIS West Africa.

Kumar, R.T., Sampson, A.-L., Dorathy, E., Wokoma, I., and Ablorh, M.A. (2013). "Study on Environmental Impact on Oil and Gas Activities in Ghana: Analysis by Graphical Approaches Using Matlab," *International Journal of Engineering Trends and Technology*, 4(3): 344–348.

Kuyini Mohammed, A. (2013). "Civic Engagement in Public Policy Making: Fad or Reality in Ghana?" *Politics and Policy*, 41(1): 117–152.

Leach, M., Mearns, R., and Scoones, I. (1999). "Environmental Entitlements: Dynamics and Institutions in Community-Based Natural Resource Management," *World Development*, 27(2): 225–247.

Lockwood, M., Davidson, J., Curtis, A., Stratford, E., and Griffith, R. (2010). "Governance Principles for Natural Resource Management," *Society and Natural Resources*, 23: 1–16.

Matheson, G. and Giroux, L. (2010). "Capacity Development for Environmental Management and Governance in the Energy Sector in Developing Countries," *OECD Environment Working Papers*, No. 25, OECD Publishing.

Mensah, M. (2010). "Challenges of Environmental Degradation—Ghana's Preparedness for Effective Oil Spill Response," *Ghana Policy Journal*, 4: 119–129.

Modern Ghana News. (2012). "Ghana's Regulatory Environmental Governance Model Gearing towards Global Benchmark," Available at: http://

www.modernghana.com/print/421187/1/ghanas-regulatory-environmental-governance-model-g.html. Accessed March 7, 2013.

Sakyi, P.A., Efavi, J.K., Atta-Peters, D., and Asare, R. (2012). "Ghana's Quest for Oil and Gas: Ecological Risks and Management Frameworks," *West African Journal of Applied Ecology*, 20(1): 57–72.

The Economist Intelligence Unit. (2013). *African Markets. Managing Natural Resources. Country Report: Ghana*, London: The Economist Intelligence Unit.

Williams, C.A. (2004). "Civil Society Initiatives and 'Soft Law' in the Oil and Gas Industry," *New York University Journal of International Law and Politics*, 36(2/3): 457–502.

World Bank. (2009). "Economy-Wide Impact of Oil Discovery in Ghana," November 30, PREM4 Africa Region.

World Bank. (2010). *Environmental Governance in Oil Producing Developing Countries: Findings from a Survey of 32 Countries*, Washington, DC: World Bank.

Yeboah, A.S., Kumi, E., and Kwarteng, E. (2012). "Empirical Assessment of Expectations Associated with the Recent Discovery of Commercialisable Oil in Ghana," *International Review of Management and Marketing*, 2(3): 177–191.

Zandvliet, R. (2013). "Civil Society and Oil and Gas Governance in Ghana," Available at: http://capacity4dev.cc.europa.eu/article/civil-society-and-oil-and-gas-governance-ghana. Accessed July 2, 2013.

8

The Capacity Question, Leadership, and Strategic Choices: Environmental Sustainability and Natural Resources Management in Africa

Korbla P. Puplampu

Context and problem

Africa has abundant natural resources, and their effective utilization is essential in enhancing the region's development conditions in the 21st century (NEPAD 2003; UNDP 2003). At the global level, the Millennium Development Goals (MDGs) include a specific goal (MDG 7) focused on environmental sustainability. What is remarkable is that the goal of environmental sustainability is integral to the attainment of all the other MDGs. For example, sustainable management of natural resources is crucial in any program to alleviate poverty (MDG 1) and childhood mortality (MDG 4) (UNEP 2007; Ochola et al. 2010; WWF and AfDB 2012). The African Union, the continental political body, has spearheaded several initiatives at the dawn of the new millennium aimed at a better understanding of the role of natural resources and environmental stewardship in the African development discourse (NEPAD 2003; African Union and NEPAD 2010).

However, African countries, in the midst of their natural resource wealth, have occupied a central place in the paradox of plenty debate, following Karl's (1997) seminal work, *The Paradox of Plenty*. The literature refers to this paradox as the natural resource "curse" and, by extension, as a "blessing" syndrome (Ross 2001, 2008; Watts 2004; Collier and Hoeffer 2005; Shaxson 2005, 2007, 2008; Humphreys et al. 2007; Stevens and Dietsche 2008; Omojola 2012). The argument relates to several factors and outcomes, beginning with how natural wealth distorts the relationship between natural resource prices and the economy, particularly the decline of the manufacturing sector (the Dutch disease). Further concerns include the decline of democratic governance, the lack of transparency, including corporate rule and elite capture of economic and political institutions, political instability, and a persistent state of armed conflict. While for several African countries, natural resource wealth has been depleted by war or given rise to prolonged and protracted conflicts, flourishing dictatorships, and the related lack of representative democracy, human rights abuses, and complete economic disaster, the resource curse argument is not without critique (Karl 1997; Davis and Tilton 2005; Mehlum et al. 2008; Obi 2010). At the heart of the critique are issues of causality, correlation, and the role of institutions. The importance of institutions, for example, stems from the fact that they offer the structure required for the effective interaction of political and economic relations, or what North (1990) calls the "rules of the game."

In Africa, as in many other parts of the world, discussions of the institutional structure inevitably lead to the state and its institutional apparatus in the development framework (Keeley and Scoones 2003; NEPAD 2003; World Bank 2004; Haslam et al. 2012). However, the role of the contemporary African state in the development process, like others, has undergone dramatic changes due to or in response to globalization. This is because with the emergence of non-state actors, including both for-profit and nonprofit institutions at the national and global levels, national development has become a contested site (World Bank 2004; Beaudet 2012). The contestation suggests that policy outcomes are not predetermined, but rather contingent on the complexities in the relationship between state and non-state actors and institutions. An essential variable in shaping policy outcomes, especially for the benefit of the society, is political leadership.

The concepts of leadership and specifically political leadership, despite their widespread interest, are illusive and contested (Cole 1994; Tettey 2012). This is largely because the concepts are examined in general terms, making for difficulties in operationalizing such general assessments. Bolden and Kirk (2009) identify four broad categories or perspectives on leadership: essentialist, relational, critical, and constructive. The essentialist perspective attributes leadership to something that leaders do, and hence depicts it in a context of "leaders" and "followers." Furthermore, good leadership is presented "as either residing in the personal qualities of the leader, the behaviours they enact and/or the functions they perform" (Bolden and Kirk 2009: 70). Leadership, in this context, is premised on Max Weber's work on formal-legal authority and the ideal-type features of bureaucracy as a formal organization (Gerth and Mills 1946). Bureaucratic institutions can best attain their goals with specialized and skilled personnel.

The relational perspectives on leadership stress the relations leaders have with others, focusing on underlying social processes and an acknowledgment of contextual and systemic forces (Bolden et al. 2008). Critical perspectives on leadership underscore the power and political aspects of institutions or organizations and argue the nonexistence of any definite account on leadership. The constructionist perspective hinges on social constructions of shared meanings in terms of how such constructions enable or provide meaning to the lives of people. Issues of process and construction loom large in this perspective (Pye 2005; Foldy et al. 2008).

The foregoing point nonetheless, as Van Wart (2005: 221) astutely notes, is that "leaders do not act [or emerge] in vacuum." Accordingly, any conceptual analysis of leadership has to pay some attention to context. One significant aspect of context for this current study is the role of leadership in the development discourse. Several, if not all, "African governments have at various times claimed that they should be judged on their ability to bring about development. Therefore, development serves as a legitimating norm for many African leaders, but it is the type of governance in play that, in large part, determines the content and direction of development" (Makinda 2012: 67).

Leadership and governance are essential in the development discourse in the African situation where there are ongoing discussions on the region's leadership and governance crisis or malaise (Obi 2001; Arriola 2009; Makinda 2012; Tettey 2012). Whatever the nature and form of leadership in vogue, the workings of the bureaucracy and leadership would inevitably emerge in terms of bureaucratic and managerial leadership (Littrell 2011; Wanasika et al. 2011). These two forms of leadership will pit public civil servants against their political masters when it comes to designing, implementing, and monitoring development policy (Owusu 2003; Booth 2011). Therefore, the definition of leadership in this study focuses on state (both appointed and elected) and nonstate officials and their capacity to forge a vision, consensus, or path for the national development effort. The capacity to forge a vision entails choices and leadership options. Nevertheless, there are structural limits on choices and political options. These complications are brought into sharp focus in an era of globalization and its implications for political, economic, cultural, environmental, and ideological relations (Held and McGrew 2004; Scholte 2005; Steger 2009).

The globalization debate is in two main categories—the hyperglobalists and global skeptics (Held and McGrew 2004; Scholte 2005; Steger 2009). Briefly stated, the hyperglobalists focus on the irrelevance of the state and privilege markets in the decision-making, choices, and options of private actors and restructured public institutions. In this argument, policy-making is reduced to invisible and inevitable processes emanating from globalization (Wolf 2005; Bhagwati 2007). Global skeptics reiterate that although the state's role is reduced, it is influential in setting the parameters for markets and private actors. The argument from this perspective is that there is an obvious place not for a strong state per se, but one that can arrive at "smart" decisions with an eye on the interests of the society (Chossudovsky 2003; Saul 2009). There is valuable empirical literature on the role of the state in an era of globalization in the African context that goes beyond the arguments of the hyperglobalists and the global skeptics (Khan 2006; Smith 2003; 2006). A key issue in the literature is the need to situate the role of the state in the complexities of state-society relations, hence, there should also be focus on domestic institutional processes (Weiss 2003). This is because policy-making in an era of globalization is a contested process between state and non-state actors. The critical issue here is the fact that the non-state actors can be located within and beyond national borders and they have differences in terms of size, motivation, and orientation. In other words, the implied minimal role of the state ceding its role of market forces is not consistent with the practical implications of the role of the state in natural resources management in Africa.

Discussions on natural resources management in Africa are not immune from forces of globalization, especially if the emphasis is on sustainable development, democratic governance, and institutional choices (Ribot 2003, 2004; Karl 2007; Mehlum et al. 2008). One aspect of globalization relevant to this study is the changing role of the state at the political and economic level and the fact that leadership and political choices will be contingent on a complex array of internal and external forces. The complexities notwithstanding, the acid test of leadership and choices are the outcome for citizens.

This chapter interrogates the relationship between leadership and strategic choices in natural resource management in Africa by addressing three related objectives. First, survey and contextualize major policies or practices of the African state in natural resource management. Second, analyze the issues and challenges surrounding the changing role of the African state in the policy framework on natural resource management. Third, discuss capacity development in terms of leadership and strategic options for sustainable development of natural resources in Africa. The remaining chapter is divided into three main sections. The section titled "Political Leadership and Strategic Choices: The Extractive Resources in Africa" examines the major policies of the African state in natural resource management. The section titled "Sustainable Development: Analysis, Issues, and Challenges" will analyze the issues and challenges stemming from the regimes of natural resource management in Africa. The final section will outline leadership and strategic options for sustainable development of natural resources in the African development agenda.

Political leadership and strategic choices: The extractive resources in Africa

The specific resources at the center of this chapter are the extractives, specifically oil, gas, and diamonds. The major African oil-producing countries affiliated with the Organization of Petroleum Exporting Countries (OPEC) are Algeria (1969), Libya (1962), Nigeria (1971), and Angola (2007). Non-OPEC African countries with considerable production potential are Chad, Gabon, Equatorial Guinea, and Sudan, and there have been recent discoveries in Ghana, Tanzania, and Uganda. Algeria and Libya have been active oil-producing countries since the 1960s. Both countries still have considerable proven oil reserves and contribute to make North Africa one of the few regions in the developing world "where oil companies have full access to [oil] reserves and/or access to reserves with the participation of the state-owned company" (Fattouch and Darbouche 2010: 1119). Abdelaziz Bouteflika has been the elected President of Algeria since 1999, and in 2009, a constitutional amendment made it possible for him to run for a third term as president. Muammar Gadhafi assumed office in Libya through a military takeover and ruled the country for 41 years until he was swept away in the popular citizens' revolt in 2011. Mohammed el-Magarief is now, following the gridlock of the general elections in September 2012, the President of Libya. Since Nigeria joined OPEC in 1971, its political leadership has shifted between military dictators (who ruled for the greater part) and constitutional democracies, especially since 1999. Currently under the leadership of an elected president, Goodluck Jonathan, Nigeria is generally regarded as a poor example of the role of oil wealth in national development (Khan 1994; Watts 2004). Under the leadership of Eduardo dos Santos, while Angola's oil wealth and membership in OPEC is relatively new, to a large extent, it mirrors the duality of wealth and misery associated with oil wealth in Nigeria (Cillers and Dietrich 2000; Le Billon 2001).

Beyond the OPEC African countries, Equatorial Guinea, Gabon, and Sudan are also major oil-producing countries, although oil production has declined in Gabon. Oil transformed Equatorial Guinea during the late 1990s "into one of the world's fastest growing economies and a sought-after political partner in the Gulf of Guinea" (Frynas 2004: 527). President Obiang Mbasogo, who deposed his uncle Macias Nguema after 11 years in power, has been at the helm of Equatorial Guinea for over three decades. In Sudan, President Omar al-Bashir has ruled for the past 20 years, and was in charge when the country began exporting crude oil in 1999 (Idahosa 2002; Obi 2007). Sudan's oil production status has been affected by the plebiscite that led to the political independence of South Sudan in 2011. Since the oil wealth is located predominantly in South Sudan, Juma (the capital of South Sudan) and Khartoum (capital of Sudan) now have to negotiate the terms of transfer of oil to ports in Sudan.

While Ghana and Uganda, as the latest oil-producing African countries, yet to be transformed in any significant way by their new source of wealth, what is essential to this study is the state of political leadership in both countries (Kathman and Shannon 2011; Kasimbazi 2012; Okpanachi and Andrews 2012). Currently, both elect their presidents through the ballot box. Yoweri Museveni has ruled Uganda for over two decades; he began as a military leader, and then later embraced democratic politics, running as a civilian; then in 2006, he abolished presidential term limits, which allowed him to run for a third term and is currently serving a fourth five-year term as president. Ghana, on the other hand, has had democratic transitions of government since the fourth Republican Constitution was adopted in 1992. The current president, John Mahama, who was elected in the December 2012 elections, was first sworn into office barely hours after his predecessor, Atta Mills, died in office in July 2012.

Diamonds are another extractive product in Africa's natural resource industry. The problem with diamonds, like any other of the natural resources in Africa and other parts of the world, is that it is a global commodity. Their marketing and subsequent consumption take place in sophisticated and distant locales, thanks to the alluring image De Beers, the South African mining giant, has fostered on the world with its "diamonds are forever" campaign. The major diamond-producing countries include Botswana, Central African Republic, Democratic Republic of Congo, Republic of Congo, Sierra Leone, and South Africa.

There is considerable consensus that Botswana, unlike other resource-rich African countries, has with sound political leadership managed its diamond and other resources in a sustainable manner (Auty 2001; Lange and Wright 2004). For Sierra Leone, in contrast, diamonds in conjunction with state collapse contributed to decades of conflict (Alao 1999; Hirsch 2001; Silberfein 2004). It is remarkable to note that the return of peace and stability in Sierra Leone in the early part of the last decade has contributed to the emergence of relatively effective institutions and economic growth, which has been driven by natural resources, including iron ore (World Bank 2012: 3). Even though the subsequent discussion will focus on the case of oil- and gas-producing countries, it is important to stress that the arguments apply to non-oil and -gas countries and hence to natural resource management in general. The aforementioned review highlights the different manifestations of political leadership or regime in resource-rich African countries, especially the longevity of most of the leaders (Table 8.1). It is important to stress that longevity of leadership does and can take place in democratic societies. Thus, the major argument Table 8.1 seeks to advance is the absence of institutional mechanisms that offer citizens the opportunity to engage in free, fair, and frequent elections and hence the selection of their political leaders.

A useful analytical framework is to focus on the interrelated themes of political leadership, policies, and institutions and the role of civil society. Political leadership sets the tone, attitude, and behavior of oil-producing countries. Take the case of resource nationalism, which is

Table 8.1   Longevity of selected African rulers and the 2012 human development status of their countries

Country	Ruler	Years in Office	UNDP 2012 Ranking
Algeria	Abdelaziz Bouteflika	13	96
Angola	José dos Santos	33	148
Cameroon	Paul Biya	30	150
Eq. Guinea	Obiang Mbasogo	33	136
Gabon	Ali Ondimba	3	106
Ghana	John Mahama	1	135
Libya	Mohammed el-Magarief	*Since Sep.* 2012	64
Nigeria	Goodluck Jonathan	2	156
Sudan	Omar al-Bashir	19	169
Uganda	Yoweri Museveni	26	161

essentially the politicization of oil markets and revenue as exemplified by members of OPEC (Stevens 2008; Bremmer and Johnston 2009; Vivoda 2009). Resource nationalism assumes that the high prices of crude oil give governments in oil-producing countries sufficient wealth that can be utilized for the national development agenda. They also use the opportunity as leverage in their relations with external actors, particularly multinational oil companies. Alternatively, downward prices compel the same governments to seek foreign investment to boost production and better manage budget shortfalls and potential deficits, sometimes by drastically loosening the regulatory framework on natural resources. This relaxed regulatory regime is not always consistent with the interests of foreign investors, especially if attempts by the government to raise further revenue decrease the profit margins of multinational oil companies.

In any case, increases in government revenue from oil wealth do not necessarily give rise to improvements in citizens' living conditions. Nigeria demonstrates how dramatic increases in oil prices do not necessarily lead to significant improvements in the lives of citizens (Obi 2007; Shaxson 2007). Despite the billions in oil revenue generated since Nigeria joined OPEC in 1971, the country on most indicators of human development has performed poorly, and its position on the human development index sustains the argument that the oil wealth has not benefited the citizens (Table 8.1). Equatorial Guinea also provides another instance of resource nationalism—in this case, in terms of relations with foreign oil companies. One remarkable aspect of the oil business in Equatorial Guinea is that the "government's venture into the oil business has been very modest, especially when compared to the achievements by other" African countries like Nigeria and Angola (Frynas 2004: 531). The underlying reason is lack of capacity, which has "forced the government to rely on various foreign advisers...and left Equatorial Guinea with less favorable deals with oil companies than most other countries in the Gulf of Guinea" (Frynas 2004: 531–532). Capacity deficiencies also account for the "discrepancies between what the companies were supposed to pay and how much they actually paid to the government," and as the government continues to place minimal fiscal and geological conditions, it is not surprising that Equatorial Guinea remains a destination of choice for multinational oil companies (Frynas 2004: 533–534). Political leadership has a role in setting the policy agenda in the midst of resource nationalism, the cyclical movement of production and price levels, and the consequent national choices. Such choices are conditioned by state ideology, the unique circumstances of the country (e.g., the colonial experience), the level of the country's socioeconomic development, and its geological characteristics (Stevens 2008). The cases of the Republic of Congo (Congo-Brazzaville) and Gabon are apposite (Shaxson 2005).

Another factor that frames political leadership and the oil and gas sector in Africa is the role of transnational oil companies based in the traditional powerful countries and regions (Europe and North America) as well as emerging development hot spots, particularly Brazil, India, and China (Alden and Davies 2006; Frynas and Paulo 2007). On the one hand, multinational oil corporations, which have no allegiance to any location and are interested only in maximizing returns on their investment, are able to set up economic activities in locales that would ensure such returns. Their economic power gives them the ability or potential to overwhelm the state, more so the African state. On the other hand, the supposedly overwhelmed African state can, by working with the broader civil society in either the national or global context, bring considerable pressure to bear on the activities of multinational organizations and influence some policy outcomes. This is because, in an era of economic globalization, "states are no longer the only significant actors...civil society advocacy groups...[also] play a major role" (Kobrin 2004: 428) in the policy arena. These dynamic relationships found tangible expression in the activities of the Canadian mining company Talisman Energy in Sudan. In October 1998, the company purchased a 25 percent stake from Arakis Energy, another Canadian company; in March 2003, it resold the 25 percent stake to a subsidiary of an Indian national oil company and pulled out of Sudan in what had by then become a controversial oil project (Idahosa 2002; BBC 2003; Kobrin 2004).

Even though Talisman's operations provided revenue that the Sudanese government used to finance the war against its own people during the conflict between northern and southern Sudan, according to Jim Buckee, Talisman's Chief Executive, the company pulled out "because of 'US Pressures' which threatened to exclude Talisman from US financial markets" (BBC 2003). While the pressure or threat by the United States to exclude Talisman from US financial markets might have made a difference in the final outcome, the clause that would have actualized the threat was eventually removed (Matthews 2004). The removal of the clause is consistent with the general stand of the United States with respect to oil-producing countries. Analysts have noted a significant policy shift in the tone and relations between the United States and oil-rich African countries, for example, Angola and Equatorial Guinea, an outcome that could well be in response to or the result of China's emerging role in Africa (Volman 2003; Abramovici 2004; Campbell 2006; Frynas and Paulo 2006; Klare and Volman 2006; Oliveira 2007; Chouala 2010).

To the United States, the conditions of access to oil constitute a national security risk; hence, the countries around the Gulf of Guinea, including Nigeria, Angola, and Equatorial Guinea, "were identified as national energy security priorities" (Chouala 2010: 146). This declaration also made it possible for several major US companies, among them ExxonMobil and ChevronTexaco, to become involved in what Ghazvinian (2007) calls the "scramble for Africa's oil." Leaders from African oil-rich countries, such as Angola's Eduardo dos Santos and Equatorial Guinea's Mbasogo, have been well received in the United States at the same time that the country showed some goodwill to Cameroon's longtime dictator Paul Biya (Chouala 2010: 157).

The final leadership variable relates to citizen engagement in terms of the extractive resources and the broader political sphere in general. Specifically, this line of discussion focuses on how citizens and society benefit from natural resource management in the case of the oil-rich African countries (Shackelton and Campbell 2000; Oyono 2004; Gibbes and Keys 2010). Table 8.1 documents how the performance of the oil-rich African countries on UNDP HDI has been nothing short of disappointment. Consequently, natural resource management in some African countries has been more of a "curse" (Sachs and Warner 2001; Humphreys et al. 2007).

The disappointment, however, has to be reconciled with specific activities of some oil-rich African political leaders. In the midst of the grinding poverty and deplorable living conditions in Equatorial Guinea, President Obiang Mbasogo decided to endow a United Nations Educational, Scientific and Cultural Organization (UNESCO) prize in his name, a proposal that deeply divided the global institution (Sharma 2010). The resultant groundswell of opposition to the initiative may have led to the name change, whereby Equatorial Guinea replaced Obiang Mbasogo. The basis for the opposition included not only the human condition of the majority of Equatorial Guineans, but the absence of democratic governance, rampant human rights violations in the country, and the perception that the prize was an attempt by the aging leader to placate international public opinion about him and, in the process, garner some international legitimacy (Sharma 2010). The controversy did not abate when the UNESCO-Equatorial Guinea International Prize for Research in the Life Sciences was finally awarded in July 2012 (Sharma 2012). A critical requirement in the management of natural resources is transparency. Chouala (2010: 156) sums up the record on transparency in the oil industry in Africa as "one of unparalleled opacity... [For instance, in] Cameroon, oil money has for a long time been excluded from public accounting by not being budgeted. In Angola, oil accounts and revenues remain 'state secrets' and there is no public transparency." In a context of political dynasties and lack of accountability, there are few tangible benefits to the society at large.

African countries are not alone in being endowed with oil wealth. One atypical global example in the management of oil wealth for national development is Norway, a small open economy in Northern Europe at the very top of the annual global index on human development (UNDP 2012). Oil was discovered in Norway in 1969, and by the 1980s, oil exports "accounted for approximately a third of Norway's exports. During the 1990s, production increased to more than three million barrels per day, making Norway for some years the world's largest exporter of crude oil" (Cappelen and Mjøset 2009: 8). There is widespread consensus in the literature on how Norway used its oil revenue to the benefit of the whole society, escaped the Dutch disease and the resource curse phenomenon, and offers a model to other oil- or natural resource-rich countries (Wigglesworth and Kennedy 2007; Velculescu 2008; Cappelen and Mjøset 2009). Beyond Norway, oil-producing African countries can also pay attention to both Venezuela and Indonesia in terms of the institutional framework in managing oil wealth for national development (Ascher 1998; Parker 2005).

Sustainable development: Analysis, issues, and challenges

Sustainable development captures the desire to use natural resources in a way that will satisfy the needs of the current and future generations. Given that natural resources constitute part of the environment, discussions on sustainable development, by definition, also entail an understanding of environmental factors. Several theoretical antecedents underpin the contemporary discussions on sustainable development. First, the path-breaking work *The Limits to Growth* by Meadows et al. (1972) made a strong case to better understand the relationship among population growth, production, and consumption because natural resources are finite. A second milestone was the United Nations-sponsored World Commission on Environment and Development in 1983. The commission, under the leadership of Gro Harlem Brundtland, produced another informative report, *Our Common Future,* which defined the concept of sustainable development as the development that "meets the needs of the present without compromising the ability of future generations to meet their own needs" (WCED 1987: 3). A third event is the 1992 Rio Earth Summit and its Agenda 21, which highlighted biodiversity as involving both being about plants, animals, and microorganisms and how these ecosystems interact with people, and their requirements for food security, medicines, fresh air and water, and a clean and healthy environment (UNCED 1993).

There is an ongoing debate on the extent to which Rio 1992, in the aftermath of 2012 Rio + 20, has attained its objectives (Beckers 2012; Moldan 2012). However, what is not in doubt is the extent to which sustainable development has compelled new ways of thinking about the relationship between the environment and economic development, thus opening up possibilities of working toward rethinking governance systems and institutions, lifestyle choices, and ideas for the future, including discussions around a green economy (Happaerts 2012; Ishwaran 2012). Therefore, sustainable development is integral to the development debate and entails the role of political leadership in managing available resources for national development. Sustainable development requires capacity, specifically the capacity "for individuals, organizations and societies to set goals and achieve them; to budget resources and use them for agreed purposes; and to manage the complex processes and interactions that typify a working political and economic system" (ACBF 2011: 30–31). Sustainable development dovetails into the broader development debate and is contingent on the role of political leadership in managing available resources for national development today and tomorrow.

Political leadership in Botswana accounts for the country's successes in managing the diamond wealth in particular and natural resources in general, since the leaders ensured that policies and institutions work. It is that capacity that has made Norway a model. As Campbell (2011) argues with reference to Norway, the sovereign wealth fund's management has been "characterized by a high degree of transparency and its managers are directly accountable to democratic institutions"; hence the continued "success has been underpinned by political will, the rule of law and developed democratic institutions." The focus then should be on the nature and role of the state bureaucracy and the overall relationship between political leaders and bureaucrats. One relevant attribute of the postcolonial African state is its "softness." The concept of the "soft" state is useful because it cuts across differentiations on the nature of the state and takes into consideration the processes that limit the power of the state (Faaland and Parkinson 1991). According to Myrdal (1968: 896), "soft" states scarcely enforce their policies. The "softness" of the state lies in its inability "to coerce people in order to implement declared policy goals... [because of] the power structure and a gap between real and professed intentions" (Streeten 1993: 1282).

At the point of political independence, the bureaucratic structure of the state administration reflected the inability to initiate and implement effective policies and hence employed clientelism and state patronage as a means of "welding fragmented and fissiparous ruling coalitions into regimes capable of maintaining a hold on state power" (Boone 1994: 110). The postcolonial bureaucracy operated in an atmosphere where the "functional notion of government... did not distinguish between decision making and implementation roles" (Chazan et al. 1992: 42). Without qualified personnel, the same bureaucrats initiated and implemented state policies, and "because of the incompetence of existing administrative agencies, political leadership respond[ed] by administrative shortcuts, and by setting up new and hopefully more responsive administrative units" (Berg 1971: 210). The implications of a "soft" state account for inadequate or ineffective development policies. To be sure, the international financial institutions, particularly the World Bank (1981), at the height of the African crisis in the 1980s, acknowledged these difficulties of the postcolonial African state. The policy recommendations under the structural adjustment, specifically the wholesale restructuring and diminishing presence of state institutions, left a capacity vacuum that has continued to affect the overall effectiveness of the postcolonial African public service (World Bank 1989).

Another issue germane to the analysis is state-society relations, specifically an active and vibrant civil society. Drawing from the case of Norway once more, the role of a vibrant civil society has been critical in the social usage of oil revenue. It is generally known that groups with a significant power base would prefer the status quo, as the cases of Nigeria and Chad attest (Obi 2007; Moss and Young 2009). There was a real desire in both countries for reforms that would deepen the role of the oil wealth in the national development process. However, in both cases, political elites basically helped themselves to the national wealth, while ordinary citizens looked on helplessly from the sidelines. In Nigeria, there was continuous bickering among government departments on the formula used to determine oil revenue disbursements to state and local governments (Ahmad and Singh 2003).

The discovery of oil in Chad set into motion a complex series of forces (Moss and Young 2009; Barma et al. 2012). First, as a landlocked country, it required a country willing to offer a point of delivery, a role that Cameroon accepted. Second, the World Bank and other financiers requested that the Government of Chad "put all direct oil revenues into an escrow account" with specific and strict requirements on spending, for example, "80% would go for direct development and poverty reduction expenditures, 10% set aside for a Future Generations Fund, 5% on the oil-producing region, and the remaining 5% for discretionary spending," and also establish a civil society watchdog (cited in Moss and Young 2009: 10). The Government of Chad accepted these requirements but, in the final analysis, failed to allocate funds toward the creation of the civil society watchdog. Chad's parliament abolished the legislation that committed the country to the Future Generations Fund and used the oil revenue to pay back the World Bank loan, after which the bank pulled out of Chad (Moss and Young 2009: 10; Barma et al. 2012: 7).

Leadership, policy choices, and the capacity question in Africa

At the dawn of the current millennium, African leaders and their international counterparts and partners recognized the need to formulate a new economic vision, which also led to new models for leadership assessment (Tettey 2012). One specific instrument is the African Union's African Peer Review Mechanism (APRM), which covers four main areas: democracy and political governance, economic governance and management, corporate governance, and socioeconomic development. A critical review of the APRM shows the mixed record of the mechanism and points to "the discomfort or disdain of many African leaders towards objective performance assessments and/or an acknowledgement of the deficit in leadership that they represent" (Tettey 2012: 41).

One major missing link in Africa's leadership deficit is a consistent place for civil society engagement in the management of natural resources for development. It is an omission that also has implications for democratic governance. That is why the vibrancy of civil society groups focusing on the oil and gas sector in Ghana holds some promise with respect to governance of the country's oil wealth (Gyampo 2011). Then again, state-society relations will require a firm institutional framework. Some global initiatives that have found expression at the country level are worth mentioning. The Kimberley Process Certification Scheme on diamonds has not been problem-free (Wright 2012). The prospects of the Extractive Industries Transparency Initiative (EITI) are tied to capacity-building elements (Ayee 2013). EITI requires transparency in accounting, and this can take place only if necessary and sufficient institutions are in place. These institutions can be established and adequately supported only if political leaders choose to provide the required leadership.

Political leaders in Chad first signed on and then reneged on establishing an institution that would be accountable for ensuring that oil revenue is actually used for national development. Nigerian leaders have presided over a situation in which oil revenue belongs to political elites alone. What is obvious in both cases is the absence of a vibrant civil society or any form of outrage from citizens. Again, leaders should seek to promote the welfare of their citizens and not divide and conquer them through identity-based political calculations. The rhetoric of development has to be replaced by a concerted and sustainable path of development. One viable path is to embrace diversification, sustainable investment, and a governance framework of the national wealth from natural resources (Okpanachi 2011). "In fact, mining value added as a percentage of GDP has actually increased in most resource-rich African countries. In Chad, Equatorial Guinea, and Sudan this share has ballooned from less than 1 percent in 1980 to 44, 92, and 15 percent, respectively in 2010" (World Bank 2012: 18). The lack of diversification or any major structural transformation of the economy is a manifestation of lack of strategic thinking and creative choices. Wealth from natural resources is finite, hence the need to invest in rainy-day funds and manage these funds with an eye on the future (World Bank 2012: 19). Sustainable environmental development should mean proper planning of revenue from the sector to shore up and transform other aspects of the political economy. If the Rio + 20 document *The Future We Want* is to have the desired impact, there clearly must be "close collaboration between the public and private sectors and the civil society" (Ishwaran 2012).

It is obvious that established democracies will not insist on democratic credentials in oil-rich African countries that offer multinational oil companies an uninterrupted flow of oil. This demonstrates the complicity of companies from the global North in the pillaging and corruption in the global South. That realization suggests that changes or attempts at sustaining democratic and economic governance will have to begin from internal social forces. This is the context for the decision

by Nigerian authorities to prosecute former US Vice-President, Dick Cheney, for bribes given to Nigerian officials when he was head of Halliburton, and the subsequent settlement with the company, illustrates the fact that corruption is not the exclusive preserve of Africans, but is largely fuelled by companies in the global North.

(Tettey 2012: 37)

The revelations that French presidents, from Charles de Gaulle to François Mitterand, sanctioned the payment of "commissions" to foreign governments that hosted French companies speak to the internal and external dimensions of corruption (de Quetteville 2001). These practices are behind the current legal troubles of Le Floch-Prigent, the former head of Elf, the French oil giant, and his subsequent extradition from Côte d'Ivoire to Togo to face fraud charges (Telegraph 2012). Part of the problem is that, so far, global governance initiatives, from either the political or economic perspective, are simply that—talk and little to no action. Booth (2011: 1) attributes the problem to how "programmes to improve governance continue to reflect what ministers and parliaments in donor countries will support, rather than a relevant body of knowledge and experience." In other words, instead of completely discarding local institutions and culture, the focus should be on exploring what aspects of the culture can be beneficial and local problem-solving should be at the center of global governance initiatives (Booth 2011). Botswana's successes stem from a careful integration of local cultural practices into the large national development conversation. While periodic and free elections are consistent with democratic practices, governance extends beyond that to include effective controls on political power.

There is no universal model to inform the interaction of political leadership, natural resources, and national development. However, to the extent that resource wealth can propel the overall development of the society, citizens should demand accountability from their leaders. This is why Moss and Young (2009) call for direct cash payments or dividends to Ghanaians. Drawing from the experiences of Alaska, the argument is that once citizens receive a direct cash benefit, it will be incumbent upon the state to institute effective taxation policies to collect the required taxes for the public good. Even though direct payments will also compel the general public to remain vigilant and demand accountability and transparency in the management of the oil wealth, questions still remain. For example, will leaders improve taxation instruments and thus tax the payments appropriately and utilize the funds for national development? In the final analysis, it is important to have not only policies and institutions, but also leaders of goodwill whose raison d'être is the national good.

References

Abramovici, P. (2004). "United States: The New Scramble for Africa," *Review of African Political Economy*, 31: 685–690.

African Capacity Building Foundation (ACBF). (2011). *African Capacity Indicators 2011—Capacity Development in Fragile States*, Harare: African Capacity Development Foundation.

African Union (AU) and NEPAD. (2010). *Africa's Capacity Development Strategic Framework*, Midrand, South Africa: NEPAD Planning and Coordinating Agency.

Ahmad, E. and Singh, R. (2003). *Political Economy of Oil-Revenue Sharing in Developing Country: Illustrations from Nigeria*, IMF Working Paper, No. 3/16, January.

Alao, A. (1999). "Diamonds Are Forever…But So Also Are Controversies: Diamonds and the Actors in Sierra Leone's Civil War," *Civil Wars*, 3(2): 47–49.

Alden, C. and Davies, M. (2006). "A Profile of the Operations of Chinese Multinationals in Africa," *South African Journal of International Affairs*, 13(1): 83–96.

Arriola, L.R. (2009). "Patronage and Political Stability in Africa," *Comparative Political Studies*, 42(10): 1339–1362.

Ascher, W. (1998). "From Oil to Timber: The Political Economy of Off-Budget Development Financing in Indonesia," *Indonesia*, 65: 37–61.

Auty, R.M. (2001). "The Political State and the Management of Rents in Capital-Surplus Economies: Botswana and Saudi Arabia," *Resources Policy*, 27(2): 77–86.

Ayee, A. (2013). "Extractive Resources Policy in Ghana," in F. Ohemeng, B. Carroll, J.A. Ayee, and A. Darku (eds.), *Public Policy Making Process in Ghana: How Politicians and Civil Servants Deal with Public Problems*, New York: Edwin Mellen Press, pp. 247–267.

Barma, N.H., Kaiser, K., Le, T.M., and Viñuela, L. (2012). *Rents to Riches: The Political Economy of Natural Resource-Led Development*, Washington, DC: World Bank.

BBC News. (2003). "Talisman Pulls Out of Sudan," Available at: http://bbc.ca.uk/pr/fr/-/2/hi/business/2835713.stm. Accessed September 20, 2012.

Beaudet, P. (2012). "Globalization and Development," in P.A. Haslam, J. Schafer, and P. Beaudet (eds.), *Introduction to International Development: Approaches, Actors, and Issues*, 2nd edition, Don Mills, ON: Oxford University Press, pp. 107–124.

Beckers, T. (2012). "After Rio + 20: From Patchy Achievements to Sustained Reform," *Environmental Development*, 3: 1–4.

Berg, E.J. (1971). "Structural Transformation versus Gradualism: Recent Economic Development in Ghana and Ivory Coast," in P. Foster and A.R. Zolberg (eds.), *Ghana and Ivory Coast—Perspectives on Modernization*, Chicago and London: University of California Press, pp. 187–230.

Bhagwati, J. (2007). *In Defense of Globalization*, New York: Oxford University Press.

Bolden, R. and Kirk, P. (2009) "African Leadership—Surfacing New Understandings through Leadership Development," *International Journal of Cross Cultural Management*, 9(1): 69–86.

Bolden, R., Petrov, G., and Gosling, J. (2008). "Tensions in Higher Education Leadership: Towards a Multi-Level Model of Leadership Practice," *Higher Education Quarterly*, 62(4): 358–376.

Boone, C. (1994). "States and Ruling Classes in Post-Colonial Africa: The Enduring Contradictions of Power," in J.S. Migdal, A. Kohli, and V. Shue (eds.), *State Power and Social Forces: Domination and Transformation in the Third World*, Cambridge: Cambridge University Press, pp. 108–140.

Booth, D. (2011). *Governance for Development in Africa: Building on What Works*, Africa Power and Politics Policy Brief No. 1, April.

Bremmer, I. and Johnston, R. (2009). "The Rise and Fall of Resource Nationalism," *Survival: Global Politics and Strategy*, 51(April–May, 2): 149–158.

Campbell, B. (2006). "Good Governance, Security and Mining in Africa," *Minerals and Energy*, 21(1): 31–44.

Campbell, J. (2011). "Norway, Nigeria and the Lessons of Oil," *Business Day Online*, Available at: http://www.businessdayonline.com/NG/index.php/analysis/commentary/29351-norway-ni. Accessed September 20, 2012.

Cappelen, A. and Mjøset, L. (2009). *Can Norway Be a Role Model for Natural Resource Abundant Countries?* United Nations University—World Institute for Development Economics Research, Research Paper No. 2009/23.

Chazan, N., Lewis, P., Mortimer, R., Rothchild, D., and Stedman, S. (1992). *Politics and Society in Contemporary Africa*, 2nd edition, Boulder, CO: Lynne Rienner.

Chossudovsky, M. (2003). *The Globalization of Poverty and the New World Order*, 2nd edition, Shanty Bay, ON: Global Outlook.

Chouala, Y.A. (2010). "Securing Access to African Oil Post-9/11: The Gulf of Guinea," in M.S. Smith (ed.), *Securing Africa: Post-9/11 Discourses on Terrorism*, Burlington, VT: Ashgate Publishing: 143–159.

Cillers, J. and Dietrich, C. (eds.) (2000). *Angola's War Economy: The Role of Oil and Diamonds*, Pretoria: ISS.

Cole, A. (1994). "Studying Political Leadership: The Case of François Mitterand," *Political Studies*, 42: 453–468.

Collier, P. and Hoeffler, A. (2005). "Resource Rents, Governance and Conflict," *Journal of Conflict Resolution*, 49(4): 625–633.

Davis, G. and Tilton, J. (2005). "The Resource Curse", *Natural Resources Forum*, 29(3): 233–242.

de Quetteville, H. (2001). "French Presidents 'Sanctioned Oil Firm Kickbacks,'" Available at: http://www.telegraph.co.uk/news/worldnews/Europe/france/1330870/French-presidents-sanctioned-oil-firm-kickbacks.html. Accessed September 26, 2012.

Faaland, J. and L. Parkinson (1991). "The Nature of the State and the Role of Government in Agricultural Development," in C.P. Timmer (ed.), *Agriculture and the State: Growth, Employment and Poverty in Developing Countries*, Ithaca, NY, and London: Cornell University Press: 247–274.

Fattouch, B. and Darbouche, H. (2010). "North African Oil and Foreign Investment in Changing Market Conditions," *Energy Policy*, 38: 1119–1129.

Foldy, G., Goldman, L., and Ospina, S. (2008). "Sensegiving and the Role of Cognitive Shifts in the Work of Leadership," *Leadership Quarterly*, 19: 514–529.

Frynas, J.G. (2004). "The Oil Boom in Equatorial Guinea," *African Affairs*, 103(413): 527–546.

Frynas, J.G. and Paulo, M. (2006). "A New Scramble for African Oil? Historical, Political, and Business Perspectives," *African Affairs*, 106(423): 229–251.

Gerth, H.H. and Wright Mills, C. (1946). *From Max Weber: Essays in Sociology*, New York: Oxford University Press.

Ghazvinian, J.H. (2007). *Untapped: The Scramble for Africa's Oil*, New York: Harcourt.

Gibbes, C. and Keys, E. (2010). "The Illusion of Equity: An Examination of Community-Based Natural Resource Management and Inequality in Africa," *Geography Compass*, 4(9): 1324–1338.

Gyampo, R. E. (2011). "Saving Ghana from Its Oil: A Critical Assessment of Preparations So Far Made," *Africa Today*, 57(4): 48–69.

Happaerts, S. (2012). "Sustainable Development and Subnational Governments: Going beyond Symbolic Politics?" *Environmental Development*, Available at: http://dx.doi.org/10.1016/j.envdev.2012.07.001. Accessed September 20, 2012.

Haslam, P.A., Schafer, J., and Beaudet, P. (eds.) (2012). *Introduction to International Development: Approaches, Actors, and Issues*, 2nd edition, Don Mills, ON: Oxford University Press.

Held, D. and McGrew, A. (eds.) (2004). *The Global Transformations Reader: An Introduction to the Globalization Debate*, 2nd edition, Cambridge: Polity Press.

Hirsch, J.L. (2001). *Sierra Leone: Diamonds and the Struggle for Democracy*, International Peace Academy, Occasional Paper Series, Boulder, CO: Lynne Rienner.

Humphreys, M., Sachs, J.D., and Stiglitz, J.E. (2007). *Escaping the Resource Curse*, New York: Columbia University Press.

Idahosa, P. (2002). "Business Ethics and Development in Conflict (Zones): The Case of Talisman," *Journal of Business Ethics*, 39(2): 227–246.

Ishwaran, N. (2012). "After Rio + 20: Translating Words in Action," *Environmental Development*, Available at :http://dx.doi.org/10.1016/j.envdev.2012.09.006. Accessed September 20, 2012.

Karl, T. (1997). *The Paradox of Plenty: Oil Booms and Petro-States*, Berkeley: University of California Press.

Kasimbazi, E.B. (2012). "Environmental Regulation of Oil and Gas Exploration and Production in Uganda," *Journal of Energy and Natural Resources Law*, 30(2): 185–221.

Kathman, J. and Shannon, M. (2011). "Oil Extraction and the Potential of Domestic Instability in Uganda," *Africa Studies Quarterly*, 12(3): 23–45.

Keeley, J. and Scoones, I. (eds.) (2003). *Understanding Environmental Policy Processes: Cases from Africa*, London: Earthscan.

Khan, S.A. (1994). *Nigeria: The Political Economy of Oil*, Oxford: Oxford University Press for Oxford Institute for Energy Studies.

Khan, S.E. (2006). "Bringing the State Back In? A Critique of Neoliberal Globalisation," *African Journal of International Affairs and Development*, 11(1): 128–144.

Klare, M. and Volman, D. (2006). "The African 'Oil Rush' and American National Security," *Third World Quarterly*, 27(4): 609–628.

Kobrin, S.J. (2004). "Oil and Politics: Talisman Energy and Sudan," *International Law and Politics*, 36: 425–456.

Lange, G.-M. and Wright, M. (2004). "Sustainable Development in Mineral Economies: The Example of Botswana," *Environment and Development Economics*, 9(4): 485–505.

Le Billon, P. (2001). "Angola's Political Economy of War: The Role of Oil and Diamonds, 1975–2000," *African Affairs*, 100: 55–80.

Littrell, R.F. (2011). "Contemporary Sub-Saharan African Managerial Leadership Research: Some Recent Empirical Studies," *Asia Pacific Journal of Business and Management*, 2(1): 65–91.

Makinda, S.M. (2012). "Africa's Leadership Malaise and the Crisis of Governance," in K.T. Hanson, G. Kararach, and T.M. Shaw (eds.), *Rethinking Development Challenges for Public Policy: Insights from Contemporary Africa*, New York: Palgrave Macmillan, pp. 54–82.

Matthews, R.O. (2004). "Canadian Corporate Responsibility in Sudan: Why Canada Backed Down," in J.J. Kirton and M.J. Trebilcock (eds.), *Hard Choices, Soft Law: Voluntary Standards in Global Trade, Environment and Social Governance*, Aldershot, UK: Ashgate Publishing, pp. 228–249.

Meadows, D., Randers, J., and Behrens, W. (1972). *The Limits to Growth*, New York: Universe Books.

Mehlum, H., Moene, K., and Torvik, R. (2008). "Institutions and the Resource Curse," in R. Congleton, A. Hillman, and K. Konrad (eds.), *40 Years of Research on Rent Seeking 2. Applications: Rent Seeking in Practice*, Berlin: Springer, pp. 245–264.

Moldan, B. (2012). "Rio, Twenty Years Later: Progress or Stagnation?" *Environmental Development*, 3: 180–181.

Moss, T. and Young, L. (2009). *Saving Ghana from Its Oil: The Case for Direct Cash Distribution*, Center for Global Development, Working Paper 186, Washington, DC.

Myrdal, G. (1968). *Asian Drama: An Inquiry into the Poverty of Nations*, New York: Twentieth Century Fund.

New Partnership for African Development (NEPAD). (2003). *New Partnership for Africa's Development—Action Plan for the Environment Initiative*, Midrand, South Africa: NEPAD.

North, D.C. (1990). *Institutions, Institutional Change, and Economic Performance*, New York: Cambridge University Press.

Obi, C.I. (2001). "Reconstructing Africa's Development in the New Millennium through NEPAD: Can African Leaders Deliver the Goods?" *African Journal of International Affairs*, 4(1 and 2): 142–175.

Obi, C.I. (2007). "Oil and Development in Africa: Some Lessons from the Oil Factor in Nigeria for the Sudan," in L. Patey (ed.), *Oil Development in Africa: Lessons for Sudan After the Comprehensive Peace Agreement*, Copenhagen: Danish Institute for International Studies (DIIS) (Report No.2007: 8): 9–34.

Obi, C.I. (2010). "Oil as the 'Curse' of Conflict in Africa: Peering through the Smoke and Mirrors," *Review of African Political Economy*, 37(126): 483–495.

Ochola, W., Sanginga, P., and Bekalo, I. (eds.) (2010). *Managing Natural Resources for Development in Africa: A Resource Book*, Nairobi: University of Nairobi Press.

Okpanachi, E. (2011). "Confronting the Governance Challenges of Developing Nigeria's Extractive Industry: Policy and Performance in the Oil and Gas Sector," *Review of Policy Research*, 28(1): 25–47.

Okpanachi, E. and Andrews, N. (2012). "Preventing the Oil 'Resource Curse' in Ghana: Lessons from Nigeria," *World Futures*, 68(6): 430–450.

Oliveira, R.S. de (2007). *Oil and Politics in the Gulf of Guinea*, New York: Columbia University Press.

Owusu, F. (2003). "Pragmatism and the Gradual Shift from Dependency to Neoliberalism: The World Bank, African Leaders and Development Policy in Africa," *World Development*, 31(10): 1655–1675.

Oyono, P.R. (2004). "One Step Forward, Two Steps Back? Paradoxes of Natural Resources Management in Cameroon," *Journal of Modern African Studies*, 42(1): 91–111.

Parker, D. (2005). "Chavez and the Search for an Alternative to Neoliberalism," *Latin America Perspectives*, 32(2): 39–50.

Pye, A. (2005). "Leadership and Organizing: Sensemaking in Action," *Leadership*, 1(1): 31–50.

Ribot, J.C. (2003). "Democratic Decentralisation of Natural Resources: Institutional Choice and Discretionary Power Transfers in Sub-Saharan Africa," *Public Administration and Development*, 23(1): 53–65.

Ribot, J.C. (2004). *Waiting for Democracy: The Politics of Choice in Natural Resource Decentralization*, Washington, DC: World Resources Institute.

Ross, M.L. (2001). "Does Oil Hinder Democracy?" *World Politics*, 53(3): 325–361.

Ross, M.L. (2008). "Blood Barrels: Why Oil Wealth Fuels Conflict," *Foreign Affairs*, 87(3): 2–8.

Sachs, J.D. and Warner. A.M. (2001). "The Curse of National Resources," *European Economic Review*, 45(4–6): 827–838.

Saul, J.R. (2009). *The Collapse of Globalism and the Reinvention of the World*, Toronto: Penguin Canada.

Scholte, J.A. (2005). *Globalization: A Critical Introduction*, 2nd edition. New York: Palgrave Macmillan.

Shackelton, S. and Campbell, B. (2000). *Empowering Communities to Manage Natural Resources: Case Studies from Southern Africa*, Pretoria, South Africa: CSIR.

Sharma, Y. (2010). "African Intellectuals Say No to UNESCO's Obiang Prize," *SciDev*, October, Available at: http://www.scidev.net. Accessed September 20, 2012.

Sharma, Y. (2012). "Controversial UNESCO Science Prize Finally Awarded," *SciDev*, July, Available at: http://www.scidev.net. Accessed September 20, 2012.

Shaxson, N. (2005). "New Approaches to Volatility: Dealing with 'Resource Curse' in Sub-Saharan Africa," *International Affairs*, 81(2): 311–324.

Shaxson, N. (2007). "Oil, Corruption and the Resource Curse," *International Affairs*, 83(6): 1123–1140.

Shaxson, N. (2008). *Poisoned Wells: The Dirty Politics of African Oil*, New York: Palgrave Macmillan.

Silberfein, M (2004). "The Geopolitics of Conflict and Diamonds in Sierra Leone," *Geopolitics*, 9(1): 213–241.

Smith, M.S. (ed.) (2003). *Globalizing Africa*, Trenton, NJ: Africa World Press.

Smith, M.S. (ed.) (2006). *Beyond the "African Tragedy": Discourses on Development and the Global Economy*, Aldershot, UK: Ashgate Publishing.

Steger, M.B. (2009). *Globalization: A Very Short Introduction*, New York: Oxford University Press.

Stevens, P. (2008). "National Oil Companies and International in the Middle East: Under the Shadow of Government and the Resource Nationalism Cycle," *Journal of World Energy Law and Business*, 1(1): 5–30.

Stevens, P. and Dietsche, E. (2008). "Resource Curse: An Analysis of Causes, Experiences and Possible Ways Forward," *Energy Policy*, 38: 56–65.

Streeten, P. (1993). "Markets and State: Against Minimalism," *World Development*, 21(8): 1281–1298.

Telegraph (2012). "Former Elf Oil Head Extradited to Togo," Available at: http://www.telegraph.co.uk/news/africaandindianocean/togo/9546309/ Former-Elf-oil-head-extradited-to-Togo.html. Accessed September 26, 2012.

Tettey, W.J. (2012). "Africa's Leadership Deficit: Exploring Pathways to Good Governance and Transformative Politics," in K.T. Hanson, G. Kararach, and T.M. Shaw (eds.), *Rethinking Development Challenges for Public Policy: Insights from Contemporary Africa*, New York: Palgrave Macmillan, pp. 18–53.

United Nations Conference on the Environment and Development (UNCED). (1993). *Agenda 21, Programme of Action on Sustainable Development—Rio Declaration on Environment and Development*, New York: United Nations Department of Public Information.

United Nations Development Programme (UNDP). (2003). *Human Development Report 2003 Millennium Development Goals: A Compact among Nations to End Human Poverty*, New York: Oxford University Press for UNDP.

United Nations Development Programme (UNDP). (2012). *Human Development Report, 2011*, New York: Oxford University Press for UNDP.

United Nations Environment Programme (UNEP). (2007). *Global Environment Outlook (GEO4) Environment for Development*, Nairobi: UNEP.

Van Wart, M. (2005). *Dynamics of Leadership in Public Service: Theory and Practice*, New York: M.E. Sharpe.

Velculescu, D. (2008). "Norway's Oil Fund Shows the Way for Wealth Funds," *IMF Survey Magazine*, July 9, Available at: http://www.imf.org/external/pubs/ft/survey/so/2008/pol070908a.htm. Accessed September 20, 2012.

Vivoda, V. (2009). "Resource Nationalism, Bargaining and International Oil Companies: Challenges and Change in the New Millennium," *New Political Economy*, 14(4): 517–534.

Volman, D. (2003). "The Bush Administration and African Oil: The Security Implications of US Energy Policy," *Review of African Political Economy*, 98: 573–584.

Wanasika, I.H, Littrell, J.P., and Dorfman P. (2011). "Managerial Leadership and Culture in Sub-Saharan Africa," *Journal of World Business*, 46(2): 234–241.

Watts, M. (2004). "Resource Curse? Governmentality, Oil and Power in the Niger Delta, Nigeria," *Geopolitics*, 9(1): 50–80.

Weiss, L. (ed.) (2003). *States in the Global Economy: Bringing Domestic Institutions Back In*, Cambridge: Cambridge University Press.

Wigglesworth, R. and Kennedy, S. (2007). "Norway Provides Model on How to Manage Oil Revenue," *New York Times*, October 17, Available at: http://www.nytimes.com/2007/10/17/business/worldbusiness/17iht-fund.4.7931109.html. Accessed September 20, 2012.

Wolf, M. (2005). *Why Globalization Works*, New Haven, CT: Yale University Press.

World Bank. (1981). *Accelerated Development in Sub-Saharan Africa: An Agenda for Action*, Washington, DC: World Bank.

World Bank. (1989). *Sub-Saharan Africa: From Crisis to Sustainable Growth. A Long-Term Perspective Study*, Washington, DC: World Bank.

World Bank. (2004). *Global Development Prospects: Realizing the Development Promise of the Doha Agenda*, Washington, DC: World Bank.

World Bank. (2012). *Africa's Pulse: An Analysis of Issues Shaping Africa's Economic Future*, October, Vol. 6, Washington, DC: World Bank (Office of the Chief Economist for the Africa Region).

World Commission on Environment and Development (WCED). (1987). *Our Common Future*, Oxford and New York: Oxford University Press.

Wright, C. (2012). "The Kimberly Process Certification Scheme: A Model Negotiation," in P. Lujala and S.A. Rustad (eds.), *High-Value Natural Resources and Peacebuilding*, London: Earthscan, pp. 181–187.

WWF and AfDB. (2012). *Africa Ecological Footprint Report*, Tunis, Tunisia: AfDB and WWF.

9
Debating Critical Issues of Green Growth and Energy in Africa: Thinking beyond Our Lifetimes

Abbi M. Kedir[1]

Introduction

From the 1960s through the 1990s, vested interests from major polluting nations and industries kept environmental movements at bay and undermined their influence in development thinking. In recent years, ecologists, some development economists, and environmental economists challenged mainstream thinking and introduced the concept of "environmentally sustainable economic growth," which is essentially "green growth" that promotes economic development through clean energy (Ekins 2000; Hahnel 2011). This paradigm shift in designing and implementing economic policies recognizes the place environment holds in the process of economic development and in writings by influential development thinkers (Collier 2010; World Bank 2010; Weigand 2011). Consequently, there is a focus on the quantitative as well as qualitative dimensions of development. There is intensified discussion on green economy among academics and policy-makers in summits and conferences, such as the December 2011 summit in Durban, the January 2012 Green Growth Knowledge Platform inaugural conference in Mexico City, and the May 2012 Global Green Growth Summit in Seoul.

The African Development Bank (AfDB) projects that every African country will be on a green growth development pathway by 2022. Encouraging measures are already under way in countries such as Ethiopia[2] (hydro); Kenya (geothermal); and Tunisia, Egypt, Mozambique, and Morocco (solar and wind). Other countries in Africa (notably South Africa and Ghana) are also developing initiatives to promote greener/clean energy development. If it can be taken as a

185

serious step toward green economy, the adoption of the Libreville Declaration by about 30 African countries demonstrates at least a political commitment.

Green growth has short-term costs with long-term gains in tackling poverty even if there is resistance to it in a fashion similar to the structural adjustment programs (SAPs) of the late 1980s (Arndt et al. 2010; Hallegatte et al. 2011; Tandon 2011; Resnick et al. 2012). Some believe that the green growth dream is not feasible and believe that the West exploits Africa's lack of expertise in green growth to switch climate aid to emission reductions and ring-fence sources of finance for development (Rogers 2010; *Development Today* 2011). Aid commentators contend that major donors might use the need to adopt green growth strategies as aid conditionality. Some point to the oversimplified contexts within which green growth is discussed and emphasize the importance of innovation as a precondition for green growth (Aghion et al. 2009). Green growth is not merely the management of natural resources; it also seeks to change the way we think about development. Green growth is one of the most challenging issues of our time, and its intergenerational implications for global development are increasingly being flagged by major development actors, policy-makers, and research centers (OECD 2011; UNEP 2011; UNESCAP 2011; AfDB 2012a; CMI 2012; World Bank 2012). The theme of the 2011 African Economic Conference was also green growth—a concept intensively discussed in the recently completed Rio + 20 conference. The long-term (2012–2022) strategy of the AfDB emphasizes green growth alongside its key message of inclusive growth (Schut et al. 2010). The World Bank believes that it is necessary, efficient, and affordable. Thus, it will not be surprising if it mainstreams green growth principles into its operations (World Bank 2012). We should note that even if Africa is not a major contributor to global climate chaos and irresponsible energy use, it might jeopardize its green economy dream by its increasingly intensive use of chemical fertilizers, which are major sources of greenhouse gas (GHG) emissions (Stern 2007). We will discuss both sides of the green growth argument along with critical issues in the context of Africa. Evidently, there are issues related to the institutional and human capacities in Africa to negotiate and deal with complex issues of climate change, design, and implementation of sustainable development projects and raising funds for a better future. Other issues include coordination with national development strategies, biofuels, fertilizers, food security, renewable energy sources, financing options, institutions, and sequencing of Africa's priorities.

The chapter attempts to provide a comprehensive and critical perspective on the current academic and policy debates on green economy. We give particular attention to case studies and issues of capacity development and policy choices (Resnick and Birner 2010). We also link green growth with agricultural transformation and food security issues. Green growth cannot be pursued without linking it with food security and fragility of states. Unchecked climate change will lead to floods, drought, and undermining of African societies. In turn, this makes relatively stable countries fragile and food insecure. Green growth attempts to contribute to solving all these complex problems we face. This study is relevant beyond the limits of our lifetime with implications for generations to come because it raises issues of economic and environmental development with a long-term perspective.

The remaining chapter is organized as follows. The section "Fundamental considerations for policy-makers" discusses that key issues are often ignored. These issues, inter alia, include the alignment/coordination of green policies with national development strategies, technology (e.g., biotechnology), infrastructure, finance, human capital, and feasibility of green growth (e.g., green revolution/agricultural transformation, food security, chemical fertilizers, biofuels). In the section titled "Complex issues," we discuss further complex issues such as financing possibilities, coordination of national development strategies with green policies, constraints to green growth, and possible solutions. Then the chapter concludes.

Fundamental considerations for policy-makers

a) Comparative advantage, political economy, and national development strategies

Africa is diverse. Most countries are dependent on agriculture, some are mineral-rich, some have fertile land, and others simply occupy vast tracts of barren dry land/desert. This diversity breeds heterogeneity in the type of environmental policies that can be pursued. Resnick (2012) pointed out the differences in three case study countries—Malawi, Mozambique, and South Africa—that have, respectively, favorable agroecological conditions, biofuel, and a coal mining industry.

A favorable condition that can be exploited is not always environmentally friendly, as in the case of coal mining in South Africa, because it is counter to the basic tenets of green growth. In fact, South Africa is one of the major contributors of GHG emissions globally and is ranked as one of the 13 largest polluters (World Bank 2011). Therefore, one challenging policy consideration is finding a green alternative that is not based on the country's comparative advantage. On the practical side, it might benefit from green trade, such as electricity imports to satisfy its high demand for electricity (Boonyasana 2012).

Malawi's agro-ecological condition is suitable for agricultural productivity expansion, but farming is based on intensive farming techniques. Use of chemical fertilizers is counter to green growth. However, if the government designs a green growth development strategy, it may lose its support base during elections due to the popularity of the fertilizer subsidy scheme. This case illustrates the complex interrelationships among African nations' comparative advantages, their current development strategies, and the political and economic consequences of transitioning to green policies. Hence, the discussion of green growth in Africa cannot be devoid of recognition of each country's comparative/ competitive advantage and considerations of alignment with national development strategies and political economy concerns (Clapp et al. 2010). Ethiopia has made notable progress in this regard with its plan for a nationwide railway transport system and the construction of one of the biggest hydroelectric dams in the continent (FDRE 2011; MOFED 2010).

Countries advance their development strategies based on their comparative advantages that are not necessarily environmentally friendly. Deviating from the potential of economies and growth linkages has detrimental political economy consequences because new development strategies are inevitably followed by distributional consequences to the majority or some strong interest groups in society. Therefore, the green growth agenda cannot be discussed separately from the comparative advantage and complex political economy issues, such as support for or opposition to a green policy initiative by the electorate. This poses the single most difficult challenge to African countries.

b) Technology implications

Technology and innovation are important elements in implementing a transition to a green economy. The technological requirements of green growth lead to drastic departures from existing ways of doing business. They can also increase imports of technologies and innovations from rich economies. This weakens the existing investment in human and physical capital in Africa, which is often aligned with past development strategies with minimal environmental concerns (Weigand 2011). One area of huge importance is thinking about issues linked with biotechnology such as environmental risk/benefit assessment of biotech products (e.g., genetically modified crops), protection of intellectual property rights for biotech inventions, and the research and development (R&D) expenditure implications and fiscal effects of biotechnology development. Green growth implies an expansion of green chemistry and biological research. From a policy and regulation perspective, the standardization of biotech products is required, as it is done for products based on fossil fuel and its derivatives. Strong institutional and legal frameworks are needed to implement green growth strategies (OECD 2012).

The question is how, in an era of austerity, African countries can afford to allocate limited resources to meet demands for new technologies. Better technological diffusion and development of innovations in poor countries is possible through foreign direct investment (Dutz and Sharma 2012). Some governments are realizing the importance of increased R&D expenditure geared toward green technology development. In the grand scheme of economic development, training for specific skills and retaining the relevant experts with competitive incentives facilitate innovation and technological absorption, which, in turn, reinforces the case for green growth.

c) Capacity building in human capital and skills

For African countries to make a transition to green economy, there needs to be the right mix of skills in the population. Skills shortage will impede transition to green economies. Even in developed countries such as Germany and the United Kingdom, there is a shortage of specialized "green" engineers. This is a daunting challenge in Africa, where the shortage extends from the high end of the skills spectrum to teachers, curriculum designers, and trainers of new skills. Bowen (2012) maintains that there are at least three ways through which the demand for skills and human capital are affected in this transition. First, there is a structural change across industries or green restructuring (such as the decline in or closure of coal mining activities). The labor intensity of a given economic activity varies by sector. For instance, most African economies are better off from an employment perspective if they focus on the labor-intensive renewable energy sector. Fossil fuel energy has a relatively high skill requirement (Pollin et al. 2009). Second, there will be new green jobs (e.g., carbon footprint assessors, biofuel crop farmers, laborers in commercial plantations). Third, the nature of existing jobs changes as they reflect energy efficiency and lower levels of application of potentially harmful technologies. The following list of measures, which are adapted from ILO and CEDEFOP (2011) and Bowen

(2012: 30), demonstrates the capacity-building requirements in relation to transition to green growth. There is a need for the following:

1. Capacity building for employers in the informal economy and micro and small enterprises to enter green markets;
2. Entrepreneurship training and business coaching for young people and adults to start up green businesses in conjunction with microfinance projects;
3. Environmental awareness among decision-makers, business leaders, administrators, and institutions of formal and informal training systems;
4. Capacity building of tripartite constituents to strengthen social dialogue mechanisms and to apply these to dialogue about accessibility to training for green jobs; and
5. Increased capacity of formal education and training systems and institutions to provide basic skills for all and to raise the skills base of the national workforce. This includes curriculum development, improving apprenticeship systems, and building synergies with nongovernmental organizations (NGOs) that provide education and training.

Institutions such as the Climate and Development Knowledge Network (CDKN) promote sustainable human development outcomes. Established in 2010, this network of think tanks works for the benefit of developing countries to design and implement climate-friendly development. It maximizes opportunities for funding, such as climate finance. There are recent initiatives in Ethiopia and other countries that fund projects aimed at building capacity for the staff of higher education institutions in the area of natural resource management and ecotourism. Another important project is the United Nations Environment Programme (UNEP)'s Capacity Development for the Clean Development Mechanism (CD4CDM). However, only a few countries are targeted by this project, such as those in North Africa.

Given its commitment to knowledge management and capacity development, the United Nations Development Programme (UNDP)'s support programs are relevant. UNDP has been supporting sustainable development initiatives and capacity development projects for green economy since the early 1990s (UNDP 2012). In Mozambique's semi-arid Guija District, a UNDP initiative is training communities to grow drought-resistant crops. Communities are trained in weather forecasts, climate information, adaptation techniques, sustainable charcoal production, and fodder production. The training led to changes in water management and pastoralist practices. Such behavioral shifts should be scaled up at the continental level.

In Zimbabwe, farmers from four villages were trained in crop mix diversification, infield rainwater harvesting, fodder conservation strategies, and production of drought-resistant crops. The lessons learned from these training programs are being disseminated through workshops, print, and radio and form part of the Climate Development Knowledge Network (CDKN). The UNDP manages the Cap-Net global network, which is a system set up for capacity building in sustainable water resources management. The network promotes resource efficiency, social inclusivity, and low carbon emissions. "Train the Trainer" programs at the local level reduce the need for external consultants. Equity and gender balance foster social inclusion, and promotion of energy efficiency and maximum use of renewable energy.

d) Infrastructure

If the physical capital accompanying growth is not of the right type and scale, green policies will not achieve the desired objectives. Building the "right" infrastructure is a serious undertaking due to the irreversible nature of such investment or the long timescale required to make changes. In Africa, there is no adequate infrastructure. This absence is a challenge as well as an opportunity. It is a challenge because most Africans do not have adequate roads, railway systems, water, sanitation, irrigation, energy, and transport services. On the other hand, the absence is an opportunity because nations have the chance to build the right type of infrastructure, one that adapts, mitigates the impact of climate change, and contributes to sustainable economic development. However, the initial investment requirements and the overall cost of building "right" are not modest (World Bank 2012). Weather and climatic conditions are very much linked to the type of infrastructure that is needed.

e) Feasibility of green growth: What about the green revolution rhetoric?

Within a framework of a specific definition of green growth in Africa, we provide critical assessment of the feasibility of green growth in Africa. Two critical elements of our focus are chemical fertilizers and biofuels.

1. Chemical fertilizers

The last two decades saw a sharp increase in the use of fertilizers in Africa. This technology was funded historically by major donors such as the World Bank, particularly in the 1970s and 1980s. The intensification of agriculture using fertilizers and high-yielding varieties is at the heart of the green revolution movement. Hence, in a bid to feed its population and increase productivity, Africa is becoming a non-negligible user of fertilizers (Morris et al. 2007). Even with fertilizers and irrigation, many African countries remain food insecure. Green economy suggests a reduced application of fertilizers, which, in turn, might exacerbate the existing food insecurity and lead to lower productivity. Therefore, green growth in agriculture might put a brake on the need to transform the agricultural sector and combat food scarcity. Many governments in Africa either subsidize (e.g., Malawi) or support a credit scheme in fertilizers (e.g., Ethiopia).

How can one reconcile fertilizer use, food security, and climate change mitigation in the context of Africa? Answering this question poses a significant challenge in a continent with high dependence on agriculture in the face of expanding population. A recent study on Kenya and Tanzania provides insight into what African policy-makers have to grapple with in a bid to make the continent food secure and, at the same time, mitigate the adverse impact of climate change. Palm et al. (2010) compared three agricultural intensification scenarios and found that at low population densities and large arable land availability, climate change mitigation and food security goals are compatible, while it is difficult to improve yields and reduce GHGs under high population densities because there is a requirement to use fertilizers more intensively.

2. Biofuels

Despite political commitments and rhetoric, the major culprits of GHG emissions—oil, gas, and coal—are still major global energy sources (Addison et al. 2011; IEA 2009). In a continent where there are 42 net oil-importing countries, secure and affordable renewable energy technologies (biofuels) are critical elements of sustainable development. However, in the recent wave of discovery of oil in some parts of Africa (e.g., Ghana and Uganda), the widespread use of renewable energy technologies is very unlikely in those countries and will take a long time. Finding alternative energy sources is of vital importance in the context of the Millennium Development Goals (MDGs), high energy prices, perpetual food crisis, and global economic slowdown. Biofuels are promoted as more environmentally friendly than fossil fuels but are acknowledged to have indirect social and economic impacts (Rogers 2010). They are produced using edible crops such as maize and soybeans, and if they are pursued as a substitute for fossil fuels, it might be counterproductive due to the likely food price hikes triggered by locking the essential crops to biofuel production. Large-scale biofuel production leads to increased demand for land, which might facilitate deforestation in the absence of ready-made arable land. Hence, biofuels might not be as environmentally friendly, and might lead to the displacement of communities to make way for foreign investors that grab land. This might have dire political consequences. The biggest problem for Africa is feeding its growing population and satisfying energy demands without compromising agricultural production systems. And growing crops for fuel seems a misplaced priority.

Collier (2010) argues that conversion of the biofuel crops into ethanol uses as much energy as it produces and is simply "an American fantasy." In Africa, ethanol is the most promising biofuel product that can be produced from different raw materials. For sustainable energy source generation and accelerated green economy via biofuels, policy-makers should consider country-specific environmental and socioeconomic conditions.

What is the state of biofuel production in Africa? It is very scant and limited in scale despite the existing huge potential. This is not a bad outcome given the difficult challenges that come with the production of biofuel crops and their conversion to fuel. For a transition at the continental level, there is a need for a coherent biofuel development strategy. Biofuel production has been started in only 13 African countries since the 1980s, is not backed by investment, is small in scale, and is hampered by poor facilities and erratic raw material supply.

Africa is at a crossroads when it comes to biofuel production using its vast arable land potential. The European Union is taking legislative measures to encourage member countries to produce and use biofuels to reduce the carbon intensity of major sectors such as transportation. However, Europe has no large unused arable land to produce biofuel crops. Hence, it targets places like Africa. Not only Europe but also emerging economies such as China attempt to secure access to huge tracts of land for biofuel crop production. China accomplished this in the Democratic Republic of Congo and Zambia; Germany in Ethiopia; and Sweden in Mozambique. Ghana, Nigeria, Tanzania, and Kenya are also involved in this trend. This development led to the recent debate on land grabbing. If policy-makers in Africa are not careful, the continent will simply be exploited to produce crops for others instead of converting them to green energy sources, and food security will suffer. If biofuel crop production by multinationals comes at the expense of dislocating crop land, Africa risks not only losing the potential to green the economies within its shores, but also the chance to reduce food insecurity and poverty. Policy-makers in Africa should make smart decisions to avoid land use patterns that have detrimental consequences. Hence, policy and regulatory frameworks are important (Amigun et al. 2011). There needs to be a serious consideration of the 2007 Addis Ababa Declaration on Sustainable Biofuels Development. The decision is not restricted to land development for biofuel crops with energy conversion in mind. It involves a difficult choice between food production and biofuel production for a greener future.

Complex issues

a) Constraints and policy options

Much of Africa suffers from high dependence on natural resources, climate vulnerability, and lack of basic infrastructure and financial and technological capacity. In such a context, it is important to restrict open access to natural resources, increase the efficiency of resource use, assess climate risk of country development strategies, and invest heavily in green infrastructure. This is the only way society can be trusted as custodians of the environment for current and future generations. Apart from infrastructure and human capacity, other constraints include the ever-expanding size of the African population, rapid urbanization, the cost of green policies, and the limited degree of political commitment to make the most of the continent's resources, such as wind and solar power. The exceptions are the North African states, such as Egypt and Morocco, which are by far more advanced in developing their renewable energy sector with clearer strategies. These countries have future ambitions even to export clean energy to Europe. Egypt established its New and Renewable Energy Authority in 1986 and is aiming to make concentrated solar power (CSP) its key green energy source. The challenge is for other African countries to make the most out of the widely available natural capital such as solar and wind energy. Undoubtedly, green energy sources are very costly, particularly in terms of high upfront fixed costs.

What is at stake with regard to solar and wind power? Africa has huge potential for solar energy due to abundant sunlight, unlike the colder and darker temperate places in the Northern Hemisphere. However, wind and solar energy sources are not as scalable in the way that nuclear power is scalable. The scale issue can be addressed if enough money is invested for research purposes.

Population growth remains one of the most potent challenges for development and environmental sustainability in Africa (Kedir 1994). In addition, the challenge to policy is to build future infrastructure based on less carbon, land, and water intensity. To circumvent the ever-expanding slum-like settlements in many big African cities, urban planners should carefully plan conducive settlement patterns that ensure a certain minimum standard of human dignity and protection in the face of natural and man-made disasters.

Despite challenges, some countries are good examples of prudent use of natural resources. This is largely based on a technology called cogeneration, which is well established in Africa (Baguant 1992). Mauritius, an extensive user of this technology, has a sugar industry that is self-sufficient in its electricity needs and efficiently uses the excess power by making it available to the national grid. By 1998, a quarter of the nation's electricity needs were met by the sugar industry, and this is expected to rise to 33 percent or more in the next few years. Karekezi (2002) maintains that modest capital investment, careful equipment selection, efficient use of energy in sugar production, and proper planning can lead to a 13-fold increase in the amount of electricity generated. Other countries such as Uganda have plans to boost their biomass energy efficiency (Kyokutamba 2012).

The main policy options for a transition to a green economy include regulation (limits on fishing, emission targets); taxation (development impact taxes, carbon tax, tax credits, and exemptions); expenditure/investment (R&D on green technologies, urban transport and dwelling infrastructure, afforestation, human capacity development); and institution (secure property rights).

b) Cost, benefits, and financing mechanisms of green growth

There is a broad consensus on the benefits of taking action to combat climate change. There are perceived multiple benefits that can be achieved at relatively low cost. The benefits are long term and nonmarket, which are difficult to quantify. The costs are short term (e.g., the upfront cost of providing green infrastructure). There is a call for action now to avoid costly retrofits in the future (UNDP 2012). Green growth has net gains in employment compared to the old paradigm of development (UNEP 2011). Social and environmental costs are reflected in prices (taxes) and subsidies. Subsidies to fossil fuel consumption in developing and emerging economies have contributed to limiting inflation and cushioning the welfare effects of global fuel price hikes. Transition to green growth suggests removal of subsidies, which converts this cost element to benefits. However, many countries are not in a position to remove subsidies on fuel and/or fertilizers. Their removal will have dire welfare consequences and lead to undesirable political dynamics, as seen in fuel price protests in major cities of Africa.

Economic modelers argue that the cost of green growth is not going to slow growth by much and provide estimates for major economies such as the United States. The existing estimates show that annual growth of gross domestic product (GDP) of the United States will fall by 0.03–0.09 percentage points for the period from 2010 to 2050. It is believed that the decline in growth for the globe (including Africa) is smaller than the above estimate for the United States (Krugman 2010). However, modeling costs and benefits of green growth are complex. For instance, no one knows the precise cost of solar energy when there is a focus on large-scale use.

c) Clean Development Mechanism (CDM)

Financial constraints are so binding that without a coordinated source of significant volume of funding, implementing the dream of green growth principles will be jeopardized. The current global financial crisis and the shrinking aid flows to Africa do not help the cause. Clean Development Mechanism (CDM) is an ambitious emission offset mechanism that will benefit Africa. The caveat with the initiative relates to its piecemeal and specific framework, which creates incentives not to reduce aggregate carbon emissions while specific emissions are avoided (Collier 2010). There are some attempts to finance projects that reduce GHGs via the CDM, such as the AfDB's Africa Carbon Support Programme (ACSP). Africa has a long way to go, because it accounts for only 2 percent of the existing CDM registered projects globally. CDM projects receive Certified Emission Reduction units (CERs) for the actual amount of GHG reduction achieved (AfDB 2010). CERs can be purchased by rich or emerging nations, which are often the major culprits of GHG emissions.

Another financial option relates to maintaining or increasing the carbon stored in forests. For instance, Reducing Emissions from Deforestation and Forest Degradation (REDD+) is a performance-based payment transfer scheme for reducing emissions from deforestation and degradation (Santos 2012). However, following a failed bid to have a global deal on REDD+ in the last Copenhagen conference, there is a raging debate about the effect of REDD+ on corruption in Africa. Thus, instead of improving forest governance and management, the initiative might be a source of environmental, social, and economic harm in the continent.

One important source of finance is the need for African institutions themselves to declare the revenue they get from natural resources and allocate them to matters of economic importance, such as infrastructure, R&D, and technology for renewable energy. This is closely linked to transparency and accountability as promoted by the Extractive Industries Transparency Initiative (EITI).

A significant proportion of Africa's financial constraints can be addressed if there are weakened incentives to plunder the continents' resources. Fund-raising from local income streams should not be patchy, but rather systematic and scaled up. The need for effective international initiatives that require transparency, especially from resource-rich countries, is urgent (Collier 2010). Local funds provide the more guaranteed form of funding source than the volatile tap of foreign aid, which can be turned off unexpectedly due to escalation of diplomatic and other rows between donor(s) and recipient(s).

d) Carbon tax, carbon limits, and Certified Emission Reduction units

Taxation is a powerful instrument and a strong disincentive to those who are responsible for GHG emissions. The need for environmental tax reform has been discussed since the 1950s. In fact, the proposal to tax those responsible for generating negative externalities dates to the 1920s, when Pigou took a position in favor of it in his book *The Economics of Welfare* (Krugman 2010). But the idea is gaining momentum now that society is faced with an impending environmental crisis, global green new deal, and carbon tax (Bovenberg and Goulder 1996; RSA 2010).

The resources needed to tackle climate change in developing countries by 2030 are estimated to range between US$140 and $175 billion (World Bank 2009). The climate adaptation cost is estimated to be an additional US$75–90 billion (Addison et al. 2011). The current Official Development Assistance ODA is far less than the amount required for climate change-related developments. Hence, carbon tax and other similar instruments are promising sources of revenue (Dervis 2008). When one can fairly calculate the social cost of emissions, it is possible to use carbon tax as an instrument, while carbon permits can be invoked when one knows the socially desirable level of quantity. Price instruments that put limits to carbon emissions can provide a cost-effective means of dealing with climate change threats.

Another instrument, conceptually equivalent to a Pigovian tax, is a system of tradable emission permits, which is often referred to as cap and trade in the literature. As a tax instrument by design, cap and trade creates a disincentive to polluters (Krugman 2010). One important aspect often ignored is the potential opposition to taxes by the taxpayer. For instance, the introduction of a carbon tax to reduce energy demand in South Africa led to increases in coal prices and real electricity tariffs. Simulation studies show that the high cost of investment in new energy-efficient technologies poses a potential harm to the growth of the economy by 2030. Thus, the introduction of a carbon tax was met with opposition, and the country faces tricky political economy complications (RSA 2011).

The market for CERs is not developed. The trading of CERs is much in doubt and insignificant with the expiration of the Kyoto Protocol at the end of 2012, and there is no clear alternative vision after the expiry date. There are other financial instruments to promote green growth. These include the African Green Fund established by Environment Africa for Southern Africa, the AfDB's African Green Fund (AfGF), and multilateral and bilateral trust funds for capacity development in environmental management (e.g., the Norwegian Trust Fund for Environmental and Social Sustainable Development and Norway's remarkable support of REDD+).

There is concern in some quarters that these funds could be dysfunctional like some of their predecessors, which were not operational but symbolic. One answer to these concerns is Botswana's Pula Fund, which is similar to Norway's Sovereign Wealth Funds of Norway: a useful funding source for future generations. There is a need to establish natural resource funds and fiscal rules to maximize the benefits accruing to countries (Humphreys et al. 2007). A notable multilateral funding source is the World Bank Climate Investment Funds. However, there is increasing criticism about their usefulness. Some NGOs argue that the fund could do more harm than good to poor nations due to its ring-fenced and conditionality-based disbursement of green funds.

Climate change mitigation and adaptation were discussed extensively in November 2011 at the 17th Conference of the Parties (COP 17) to the United Nations Framework Convention on Climate Change (UNFCCC) and within the Kyoto Protocol. To mitigate the impact of climate change and adapt to new circumstances, financing is paramount. The AfDB is supporting small and medium entrepreneurs in Africa via the Sustainable Energy Fund for Africa (SEFA), which is funded by Denmark and is expected to develop into a multi-donor fund. In addition, the

Scaling Up Renewable Energy Program in Low Income Countries (SREP) supports the development of pilot renewable energy strategies in the context of poverty reduction in Ethiopia, Kenya, and Mali. Similar initiatives can target other countries. For example, Angola has huge potential of hydroelectric power, while South Africa has potential for solar power. This is one of the programs under the Climate Investment Funds' (CIFs) Strategic Climate Fund (SCF), which helped to kick-start a number of large-scale clean projects. The AfDB funds green growth-targeted initiatives. It supports energy, transport, and other sectors in Africa that promote clean energy solutions using a Clean Technology Fund (CTF) to the tune of $625 million (AfDB 2010).

Approved CTFs so far are concentrated in countries located in the north of the continent, the large and richer sub-Saharan nations and also in South Africa. Plans are approved to develop renewable energy sources such as wind (Egypt, Morocco,), solar (MENA region), and energy conservation (in urban transport in Morocco and Nigeria). There is a significant funding gap for the rest of Africa. There is a need for aggressive and genuine continued commitment from the AfDB along with the other sources identified above. It requires efforts within Africa and, more importantly, a commitment for a coordinated global donor action with a strong backing for funding (Sandler 2004).

There is scope for private sector financing, which is critically important. In Africa, the private sector can be an important source of green jobs by focusing on environmentally friendly enterprise development. Private actors can work on a range of green projects, such as recycling, waste management, and renewable energy product development. Governments can also put in place effective regulatory procedures to let the private sector get green loans, grants, and seed funds.

e) Regional and global solutions

The green growth agenda has multilayered and complex national, regional, and global dimensions. Implementation of multi-country declarations and conventions that are in place is one of the biggest challenges. The Libreville Declaration on Health and Environment (2008) and the Libreville Declaration on Biodiversity and Poverty Alleviation in Africa (2010) are political commitments of many nations to move toward a green economy. Both encompass the principles of the Rio 1992 and Rio + 20 declarations on environment and development.

There are calls for electricity trading to reduce carbon dioxide (CO_2) emissions and global warming. Regional integration and global trading arrangements can facilitate this exchange. Using insights from international trade theory, Boonyasana (2012) rigorously examined whether international cooperation (such as electricity import and export) can reduce CO_2 emission levels. The panel data analysis covers 131 countries and also divisions of countries by continent, with yearly samples for the period 1971–2007. The results show that electricity cooperation is highly significant in decreasing CO_2 emissions per unit of generation. At the continent level, Asia shows the highest decline in CO_2 emissions from electricity import, with the lowest decline being for Africa due to a number of barriers to electricity trading. This rigorous and promising study reveals that electricity cooperation can have a positive impact on the efficient management of environment and the decarbonization of energy supply, and serve as an instrument for governments in the fight against global warming. If more countries become involved in electricity trading/cooperation, our planet's burden of CO_2 emissions will decrease. This is one of the complex cases of global collective action that can only be handled in multinational forums (Sandler 2004).

f) Mismatch between green growth and existing development strategies

The development strategies of many countries around the globe (not only those in Africa) are at odds with green growth principles. That is why policy-makers' and development practitioners' engagement with green growth represents a paradigm shift in thinking and practice. The key contribution of opinion formers, academic researchers, and institutions such as the African Capacity Building Foundation is to draw lessons on sequencing, coordinating, and exploiting the potential synergies of existing development policies of African states and show how they can be developed to incorporate the principles of green growth. This is a huge, ambitious undertaking that needs to be carried out urgently by all African countries to avert a mismatch between existing policies and the future of green-oriented growth.

Globally, multilateral organizations such as the World Bank that play a big role in national policy-making should integrate green growth principles into their operations. The expectation is that this will happen soon, as the World Bank has pushed the green growth agenda in different forums in recent years. For instance, the World Bank, supported by a new Green Growth Trust Fund financed by the Korean government, has recently been working on green growth issues with the aim of incorporating them into project design; technical assistance; and country, regional, and sectoral strategies such as agriculture, urban infrastructure, and transport. This will be the right move for more rapid integration of national economic policies, poverty reduction strategy papers, concentrated solar power, and green growth ideals. The same integration effort should be pursued by regional development institutions without creating inconsistency with the global-level effort to bring green growth and national strategies together.

Conclusion

If we care for our children and think beyond our lifetimes, policy-making should be honest, discourage corruption, and curb the plunder of the continent's natural resources in Africa. There needs to be a careful coordination between green growth and green revolution principles. The obvious impediments are a shortage of skills, technology, infrastructure, and finance needed to transform the economies along a green trajectory. More important, most discussions do not focus on the comparative/competitive advantages of countries and lack appreciation of the heterogeneity of green growth policies within the continent. Many institutions and organizations are still working following the old development paradigm. There is growing grassroots frustration due to the inaction by policy-makers (Bond 2011; Bond and Desai 2011).

There are also few political economy considerations in the discussions. More worryingly, African countries' green growth strategies neglect existing national development strategies developed based on comparative advantage positions that are not necessarily environmentally friendly. There are multilateral declarations (e.g., the Abuja declaration of 2012–2015 on fertilizers) that are at odds with the green growth agenda. The world continues to struggle to implement the decisions of the United Nations Conference on Environment and Development (UNCED) and Rio + 20 (Puppim de Oliveira 2012). Hence, there needs to be coordination of policies (green growth with green revolution; green growth with Poverty Reduction Strategy Papers (PRSPs); Country Strategy Papers (CSPs), since green growth strategies that create standalone documents might lead to parallel dysfunctional and uncoordinated economic policies. At a global level, recognition of Africa's position in geopolitical terms is crucial for any global deals that have an impact on the environment, and, in turn, on the continent's green development agenda.

Energy self-sufficiency initiatives such as biofuel crop production and conversion to ethanol pose a tricky challenge for Africa. The continent's nations are far behind in terms of integrating all these issues and coming up with a clear, workable green growth strategy. However, there are encouraging developments, such as the renewable energy initiatives in Morocco, Egypt, and other North African states along with cogeneration pioneers such as Mauritius, that need to be scaled up at the continental level.

Green growth is a long-term goal. Obstacles are political, behavioral, and financial. The future solution lies in an interdisciplinary approach of economics, law, engineering, political science, and social psychology. A complex interplay of political, economic, and social objectives must be evaluated in an African context as well as take into consideration the continent's often contentious and unequal interaction with others in a global setting. Different countries can have different green growth models. For instance, wind technology might be appropriate in coastal Africa; not all countries have hydropower potential; and for most of Africa, solar energy is a reasonably viable long-term option. With expectations that rich nations make clean technology available to poor countries, a realistic goal is for African governments to take the initiative to commit to green regulations, institutions, and policies and collaborate with the private sector to transform economies along green pathways (Aghion et al. 2009).

Notes

1. I am grateful to the insightful discussions I had with Adam Elhiraika, Marianna Maculan, and George Kararach. All errors and omissions are mine.
2. Ethiopia prepared a green economy strategy that aims to achieve transformation to middle-income country status by 2025 (FDRE 2011). Its commitment to environmentally sustainable development activities such as biofuel production, efficient use of forest resources, and adaptation to climate change are also highlighted in its Growth and Transformation Plan (MOFED 2010).

References

Addison, T., Arndt, C., and Tarp, F. (2011). "The Triple Crisis and the Global Aid Architecture," *African Development Review*, 23(4): 461–478.

AfDB. (2010). Climate Finance Newsletter, December, Tunis.

AfDB. (2012a). *The Long Term Strategy 2013–2022*, Tunis, Tunisia: African Development Bank.

Aghion, P., Hemous, D., and Veugelers, R. (2009). "No Green Growth without Innovation," Bruege Policy Briefs No. 2009/07, November.

Amigun, B., Musango, J., and Stafford, W. (2011). "Biofuels and Sustainability in Africa," *Renewable and Sustainable Energy Reviews*, 15: 1360–1372.

Arndt, C., Benfica, B., Tarp, F., Thurlow, J., and Uaiene, R. (2010). "Biofuels, Poverty, and Growth: A Computable General Equilibrium Analysis of Mozambique," *Environment and Development Economics*, 15(1): 81–105.

Baguant, J. (1992). "The Case of Mauritius," in M.R. Bhagavan and Karekezi, S. (eds.), *Energy Management in Africa*, London: ZED books and African Energy Policy Research Network (AFREPREN).

Bond, P. (2011). *Politics of Climate Justice: Paralysis Above, Movement Below*, Pietermaritzburg, South Africa: University of KwaZulu-Natal Press.

Bond, P. and Desai, A. (2011). *Durban's Climate Gamble: Trading Carbon, Betting the Earth*, Pretoria, South Africa: University of South Africa Press.

Boonyasana, K. (2012). "World Electricity Cooperation," Unpublished PhD thesis, University of Leicester, UK.

Bovenberg, A. and Goulder, L. (1996). "Optimal Environmental Taxation in the Presence of Other Taxes: General Equilibrium Analysis," *American Economic Review*, 86(4): 985–1000.

Bowen, A. (2012). "'Green' Growth, 'Green' Jobs and Labor Markets," Policy Research Working Paper 5990, World Bank, Washington, DC.

Clapp, C., Briner, C., and Karousakis, K. (2010). *Low Emission Development Strategies: Technical, Institutional and Policy Lessons*, Paris: OECD/IEA.

CMI (Centre for Mediterranean Integration). (2012). *Toward Green Growth in Mediterranean Countries: Implementing Policies to Enhance Productivity of Natural Assets*, Marseille, France: CMI.

Collier, P. (2010). *The Plundered Planet: How to Reconcile Prosperity with Nature*, London, UK: Penguin Books.

Dervis, K. (2008). *The Climate Change Challenge, WIDER Annual Lecture 11*, Helsinki: UNU-WIDER.

Development Today. (2011). "Bypassing Africa: Donors Pour Climate Aid into Emissions Reductions," 17: 7.

Dutz, M.A. and Sharma, S. (2012). "Green Growth, Technology and Innovation," Policy Research Working Paper 5932, World Bank, Washington, DC.

Ekins, P. (2000). *Economic Growth and Environmental Sustainability: The Prospects for Green Growth*, London and New York: Routledge.

FDRE. (2011). *Ethiopia's Climate Resilient Green Economy: Green Economy Strategy*, Addis Ababa: Ministry of Finance and Economic Development.

Hahnel, R. (2011). *Green Economics: Confronting the Ecological Crisis*, New York and London: M.E. Sharpe.

Hallegatte, S., Heal, G., Fay, M., and Treguer, D. (2011). "From Growth to Green Growth: A Framework," Policy Research Working Paper 5872, World Bank, Washington, DC.

Humphreys, M., Sachs, J., and Stiglitz, J. (2007). *Escaping the Resource Curse*, New York: Columbia University Press.

IEA. (2009). *Key World Energy Statistics 2009*, Paris: International Energy Agency.

ILO and CEDEFOP. (2011). *Skills for Green Jobs: A Global View*, Geneva: ILO.

Kammen, D., Kapadia, K., and Fripp, M. (2004). *Putting Renewables to Work: How Many Jobs Can the Clean Energy Industry Generate? Report of the Renewable and Appropriate Energy Laboratory*, University of California, Berkeley, April.

Karekezi, S. (2002). "Renewables in Africa—Meeting the Energy Needs of the Poor," *Energy Policy*, 30: 1059–1069.

Kedir, A. (1994). "The Nexus among Population Growth, Environmental Degradation and Agricultural Productivity," *Proceedings of the 4th Annual Conference*, The Ethiopian Economic Association, Addis Ababa.

Krugman, P. (2010). "Building a Green Economy," *The New York Times Reprints*, New York.

Kyokutamba, J. (2012). "Potential of Biomass Based Cogeneration in Uganda Compiled," Working Paper 394, AFREPREN/FWD, Nairobi.

MOFED. (2010). *Growth and Transformation Plan*, Ministry of Finance and Economic Development, Addis Ababa.

Morris, M., Kelly, V., Kopicki, R., and Byerlee, D. (2007). *Fertiliser Use in African Agriculture: Lessons Learned and Good Practice Guidelines*, Washington, DC: World Bank.

OECD. (2011). *Towards Green Growth*, Paris.

OECD. (2012). "Biotechnology Update: Internal Coordination Group for Biotechnology (ICGB)," OECD Newsletter, No. 23, Paris.

Palm, C., Smuklera, S., Sullivana, C., Mutuoa, P., Nyadzia, G., and Walsh, M. (2010). "Identifying Potential Synergies and Trade-Offs for Meeting Food Security and Climate Change Objectives in Sub-Saharan Africa," *Proceedings of Academy of Social Sciences*, early edition, (Special Feature): 1–6.

Pollin, R., Heintz, J., and Garrett-Peltier, H. (2009). *The Economic Benefits of Investing in Clean Energy*, Department of Economics and Political Economy Research Institute (PERI) University of Massachusetts, Amherst, USA.

Puppim de Oliveira, J. (2012). *Green Economy and Good Governance for Sustainable Development: Opportunities, Promises and Concerns*, Tokyo, Japan United Nations University Press.

Republic of South Africa (RSA). (2010). *Reducing Greenhouse Gas Emissions: The Carbon Tax Option*, Pretoria, South Africa: National Treasury, Government of the Republic of South Africa.

Republic of South Africa (RSA). (2011). *Integrated Resource Plan for Electricity: 2010–2030* (Revision 2 Final Report), Pretoria, South Africa: Department of Energy, Government of the Republic of South Africa.

Resnick, D. and Birner, R. (2010). "Agricultural Strategy Development in West Africa: The False Promise of Participation," *Development Policy Review*, 28(1): 97–115.

Resnick, D., Tarp, F., and Thurlow, J. (2012). *The Political Economy of Green Growth*, UNU-WIDER, Working Paper No. 2012/11, Helsinki.

Rogers, H. (2010). *Green Gone Wrong: How Our Economy Is Undermining the Environmental Revolution*, London and New York: Verso.

Sandler, T. (2004). *Global Collective Action*, Cambridge, Cambridge University Press.

Santos, A. (2012). *Transition to a Greener Economy in Africa*, Mozambique National Office, Mozambique, African Development Bank.

Schut, M., Bos, S., Machuama, L., and Slingerland, M. (2010). *Working towards Sustainability: Learning Experiences for Sustainable Biofuel Strategies in Mozambique*, Wageningen, the Netherlands: Wageningen University and Research Centre.

Stern, N. (2007). *The Economics of Climate Change: The Stern Review*, Cambridge, UK: Cambridge University Press.

Tandon, Y. (2011). "Kleptocratic Capitalism, Climate Finance, and the Green Economy in Africa," *Capitalism Nature Socialism*, 22(4): 136–144.

UNDP. (2012). *Comparative Experience: Examples of Inclusive Green Economy Approaches in UNDP's Support to Countries*, Environment and Energy Group, New York, USA.

UNEP. (2011). *Towards a Green Economy: Pathways to Sustainable Development and Poverty Eradication—A Synthesis for Policymakers*, Nairobi: United Nations Environment Programme.

UNESCAP (United Nations Economic and Social Commission for Asia and the Pacific). (2011). *What Is Green Growth?* UNESCAP: Bangkok.

Weigand, M. (2011). "Green Growth: Emerging Paradigm Shift in Development and Economic Growth," *Asia-Pacific Business and Technology Report*, 3(8): 8–12.

World Bank. (2009). *World Development Report 2010: Development and Climate Change*, World Bank, Washington, DC.

World Bank. (2010). *Development and Climate Change*, World Development Report, World Bank, Washington, DC.

World Bank. (2011). *World Development Indicators Database*, World Bank, Washington, DC.

World Bank. (2012). *Inclusive Green Growth: The Pathway to Sustainable Development*, World Bank, Washington, DC.

10

Moving Africa beyond the Resource Curse: Defining the "Good-Fit" Approach Imperative in Natural Resource Management and Identifying the Capacity Needs

Francis Owusu, Cristina D'Alessandro, and Kobena T. Hanson

Introduction

"Resource curse"—a paradoxical situation in which countries with an abundance of nonrenewable resources experience stagnant growth or even economic contraction—has become an important cautionary concept in discussing potential scenarios for the recent natural resource-led development efforts in Africa (Barma et al. 2012). The historical and contemporary records of many natural resource-dependent African countries justify the concerns that have been evoked in the resource curse discussions (Besada 2013). However, the need for caution inherent in policies aimed at avoiding resource curse has often led to timid (conservative) policies with some important unintended consequences. For instance, well-intentioned stakeholders in the natural resource value chain (e.g., extractive industry developers; development partners such as the World Bank and International Monetary Fund (IMF); nongovernmental organizations (NGOs) such as Revenue Watch and Oxfam International; and civil society organizations) have pushed African countries to implement "best practice" policies such as saving abroad in the form of sovereign wealth funds, oil stabilization funds, etc., as strategies to avoid the resource curse. However, given that the rate of private return on investment in Africa is higher than in any other region (Collier and Warnholz 2009), saving abroad could have a stifling effect on the development aspirations of African countries than what the resource curse itself could entail.

Given the development challenges facing many resource-endowed African countries, the threat of a resource curse and its impact on their development aspirations cannot be underestimated. The question is whether there are alternative policies that better reflect the challenges faced by different African countries. This concluding chapter argues that the resource curse is not, and should not be seen as, inevitable for resource-rich African countries. Instead of the one-size-fits-all solutions that this "best practice" approach often engenders, we propose a political economy or "good-fit" approach that takes into consideration the political economy context of individual countries in their attempt to avoid the resource curse. This requires an assessment of each country's political economy and institutional environment as it relates to natural resource management, including the considerations of the type of natural resource available and its spatial distribution; the different stages of the natural resource value chain; the political, economic, and institutional capability of the country; and the different stakeholders involved in different ways and at different levels. We argue that the institutional quality and capacity of countries are critical factors in determining the developmental outcomes associated with resource dependence. Thus, for African countries to benefit from natural resource endowments, they will require effective governance, leadership, and superior capacity of all the actors and stakeholders involved in natural resource extraction, processing, marketing, and management of revenue.

The rest of the chapter is divided into five sections. The next section briefly discusses the resource curse concept and argues that the resource curse is not inevitable. We then make a case for a political economy approach to natural resource management that stresses the need for "good-fit" policies rather than "best practices." This is followed by outlining some prerequisites for implementing the good-fit approach, including the need for interdependence, coordination, and mutual monitoring among stakeholders; creative ways for handling resource rents; and the need for alternative natural resources governance that gives priority to local communities. We then highlight the critical roles of capacity, leadership, and governance in assessing the local political and economic context and in designing and implementing good-fit policies. We conclude in the final section.

Resource curse is not inevitable

The resource curse argument is based on the relationship between resource abundance and economic growth (Corden and Neary 1982;

Sachs and Warner 1995, 2001). One would expect abundant natural resources to play a positive role in promoting economic development in poor countries because natural resources can give an economy a "big push" via initial and continued investment, facilitate skills transfer and infrastructure development, and generate employment and revenue for both central and local governments. However, many researchers have found evidence to the contrary—that is, resource-rich states tend to grow less rapidly than resource-scarce states. For instance, Sachs and Warner (1995) examined the relationship between economic growth and natural resource endowment and concluded that, when differences in macroeconomic policies and initial income levels were accounted for and adjusted, resource-rich developing states were, on average, destined to experience slower growth. This situation has been referred to as the "resource curse" (see also Strauss 2000; Gylfason 2004; Mehlum et al. 2006).

Various explanations have been given for the disappointing growth performance of resource-rich countries (Bulte et al. 2005). One explanation is the Dutch disease model, which assumes that manufacturing is the main driver of the economy and argues that a resource boom will divert a country's resources and efforts away from activities that are more conducive to long-run growth. Another explanation of the resource curse draws from rent-seeking models. The argument here is that resource rents are easily appropriable, and create the potential for bribes, distortions in public policies, and diversion of labor away from productive activities and toward seeking public favors. They also slow down institutional capacity-building efforts of the country. A third group uses institutional explanations to account for the resource curse. Like the rent-seeking models, proponents focus on the connections between resources and institutions; however, they add that the type of resource matters and also regard the form of government (and its policies) as the salient institutional feature[1] (Auty 2001). Unlike the Dutch disease and rent-seeking models, which seek to identify the best approaches for avoiding the resource curse, the implication of the institutional explanation is that the resource curse is not imperative; rather, the nature of the resource, and the quality of institutions in a country, including their capacities and the ability to formulate and implement sound macroeconomic policies, are critical in transforming the natural resource potential into a sustainable development outcome.

More importantly, the institutional explanation provides the context for understanding Africa's paradoxical situation. The region has been blessed with some of the most sought-after natural resources in the world, including massive amounts of oil reserves, minerals (including huge deposits of diamonds, phosphate, and cobalt reserves), large expanse of potential land for rain-fed agriculture, and many transboundary rivers and lake basins, yet the continent has been unable to transform its enormous economic potential and wealth into tangible benefits in terms of human security, sustainable peace, and development (Besada 2013). However, given the differences in resource endowments, as well as the economic and political contexts of the countries of the region, converting natural resource wealth into development will require more focused and prioritized approaches that attend to the near-term capacity challenges in some countries (e.g., conflict-affected and fragile countries), and the deeper institutional capacity development in others (see ACBF 2013).

Political economy approach to natural resource management

The recognition of the importance of national political and economic context in natural resource management has led to calls to de-emphasize the so-called best practices and promote "good-fit" policies in resource-rich poor countries. Generally referred to as the political economy approach to natural resource management, proponents of the good-fit policies argue that welfare-promoting policies, institutions, and governance must be tailored, at least in part, to suit each country's specific context (Barma et al. 2012; Atinc 2013; Diamond and Mosbacher 2013).

Note that the "good-fit" policies approach is gaining increased support in other areas in the development literature, including governance. One of the leading proponents of the good-fit approach in the governance literature is Grindle (2004, 2007), who criticized the push for good governance in poor countries and made the case for "good enough governance," which she defined as "a condition of minimally acceptable government performance and civil society engagement that does not significantly hinder economic and political development and that permits poverty reduction initiatives to go forward" (Grindle 2004: 526).

Grindle and the other proponents of good enough governance approach criticize the good governance idea as an "ideal state" of governance that can be reached by only a handful of countries. In addition, good governance policy prescriptions are seen as doctrinaire, unrealistic, and externally imposed. Instead, they argue that to succeed, good governance agenda must be made more feasible based on the recognition that countries differ significantly in their capacities and in the interest of their political leaders in governance reforms. The good enough governance idea also requires a more engaged civil society, with roles that go beyond mere consultations. Civil society is expected to be an important stakeholder of state governance and ensure the delivery of key services. To this end, some countries have created public oversight bodies made up of representatives of civil society organizations (Oshewolo and Oniemola 2011). In sum, the good-fit or political economy approach to natural resource management has a lot in common with the good enough governance approach.

Barma et al.'s (2012) articulation of the political economy approach to natural resource management is useful for our purpose. The main argument of these authors is that policies will be effective in leveraging natural resource-led development only when such policies are compatible with the level of institutional quality and the political economy context of the country in question. The type of the natural resource available to the country as well as the national context in the country play critical roles in this process of transforming natural resource potential into sustainable development outcomes. Based on this, the authors argue that designing natural resource management policies for any country should be preceded by an assessment of that country's political economy and institutional environment as it relates to natural resource management and, on that basis, develop a "set of targeted prescriptions across the natural resource value chain that are technically sound *and* compatible with the identified underlying incentives" (Barma et al. 2012: 3, emphasis in original). In sum, like the other proponents of the institutional explanation, Barma et al. (2012) make a case for good-fit policies for resource-rich African countries. They however go a step further and argue that identifying what the good-fit policies are requires a good understanding of the national context. It is this addition of an analytical framework to policy dimensions of the institutional explanations of the resource curse that is significant here.

Assessing a country's political economy and institutional environment as it relates to natural resource management involves considerations of the type of natural resource available, including its spatial distribution; the different stages of the natural resource value chain; the political, economic, and institutional capability of the country; and the different stakeholders involved in different ways and at different levels (Barma et al. 2012). The type of natural resources available in the country in question shapes the political economy context and conditions its overall development process. For example, resource rents from hydrocarbons and minerals influence the political economy of resource-dependent countries. Extractive industries also tend to encourage a focus on short time horizons and the pursuit of private enrichment over public welfare (Barma et al. 2012). These challenges however tend to vary depending on the nature of the dominant resources. For instance, hydrocarbons-dependent countries (oil and gas) can expect different sets of challenges from minerals-dependent economies (Barma et al. 2012). Natural resource management value chain is also a useful framework for assessing the governance and political economy parameters that affect a country's ability to transform the natural resource potential into sustainable development (Barma et al. 2012). Countries face different challenges as they proceed along the natural resource value chain, including their ability to include dynamic feedback loops in their decision-making process.[2] Finally, best-fit policies must acknowledge the enormous variations among the resource-endowed African countries in terms of the political and economic conditions, institutional quality, and capacities. For instance, the natural resource management challenge will vary for fragile/failed states such as South Sudan, for post-conflict states such as Sierra Leone, as well as for stable states such as Botswana (Besada 2013).

In sum, variations in resource-rich African countries in terms of the type of natural resource available; the different stages of the natural resource value chain; the political, economic, and institutional capability of the country; and the different stakeholders involved in different ways and at different levels suggest that not all resource-endowed countries can implement the one-size-fits-all solutions of the "best practice" approach. Rather, these variations point to the need for good-fit policies, with tailor-made solutions/strategies, based on the assessment of the country's political economy. This also calls for a sequencing of any capacity development efforts and/or rollout of policies. Depending on country-specific contexts, it may be meaningful in one instance to pursue capacity development in institutions first, as opposed to individuals. In other situations, it will be better to invest first in the capacity of the legislature, as opposed to say capacity of the government or civil society (ACBF 2013).

Prerequisites for implementing the good-fit approach

We have argued that the best practice approach to natural resource management has led to one-size-fits-all strategies that have ignored national contexts, and have therefore failed to take advantage of distinct policy priorities and reform opportunities in particular countries.

The good-fit approach overcomes this by attempting to narrow "the gap between expectation and reality with regard to interventions, aiming to deliver improved outcomes through incentive-compatible entry points and institutional designs" (Barma et al. 2012: 220). Implementing the good-fit approach therefore involves tailoring interventions to the national context and developing an incentive structure that supports and nudges stakeholders into making developmentally oriented decisions. The importance of governance, leadership, capacity, and stakeholder engagement in this effort cannot be overemphasized (see Chapters 3 and 8). Yet it is important to echo Diamond and Mosbacher's (2013) sentiment about the leadership of African governments in this process without downplaying the role of external stakeholders in the effort: "oil-rich developing countries that want to avoid the resource curse cannot wait for the international system to fight corruption for them." Governments can also enhance their ability to harness natural resource wealth for development if they make "credible intertemporal commitments to both extractive companies and its own citizens, and when the political regime is inclusive" (Barma et al. 2012: 12). The role of civil society in fighting corruption and ensuring dialogue among the various stakeholders is also crucial. In the following subsections, we outline some of the changes we envisage in this process.

a) Changing roles of stakeholders: Interdependence, coordination, and mutual monitoring

The establishment of new players such as China, Brazil, and other countries as major actors in the global resource market is generating competition for Africa's resources and changing the way they are managed. African governments can take advantage of this opportunity in pursuing their developmental objectives. As Besada (2013: 19–20) argues, the entry of the emerging economies in the global resource market has leveled

> the playing field for host countries that are negotiating contracts, pertaining to natural resources, with Western traditional powers. If these relationships are managed properly, they could be the key that African countries require for achieving sustainable economic development…Demand from the emerging powers is driving prices of mineral resources to record highs. This is a moment when the market is providing an opportunity for the continent to prosper on more equal terms.

What is required here is for African countries to claim a central role in negotiations at all levels of the natural resource value chain. While not all African countries have the needed technical capacity and necessary institutions for achieving such objectives, this is an empirical concern that cannot be assumed a priori. International actors can certainly nudge governments that are unwilling to do this; such influence on policy-makers in resource-dependent countries, especially those characterized by poor quality governance, is often limited. Even for such countries, the need of initial capital, the need for support in times of adverse shocks, and the need for validation from the international community can provide some windows of opportunities for such interventions. On the other hand, civil society, if provided with the needed capacities and with the minimum level of funding opportunities to function, can contribute greatly to this multi-stakeholder dialogue, ensuring a better policy-making for natural resource management. In sum, an understanding of the roles and capacities of the various stakeholders in the country and a renegotiation of their mandates with an eye toward more interdependence, coordination, and mutual monitoring must precede the good-fit approach.

b) Defying the Dutch disease: Investing to invest

One of the thorny issues facing many resource-rich African countries involves using funds from natural resources during the boom years for domestic investment. This is what Collier (2010) describes as the final and the weakest link in the chain of decisions of harnessing natural assets for sustained development. Many development experts, including the IMF, draw on the Norwegian experience to warn African countries to limit domestic investment due to the "absorption" problem of the economies—that is, where the economy cannot absorb the extra spending. The argument is that due to their structural problems, including inefficient bureaucracies, shortage of skilled workers, and inadequate infrastructure that are incapable of dealing with the massive revenue inflows and the expansionary public sector programs, poor countries cannot properly absorb resource rents and will therefore experience a decline in the benefits (Van der Ploeg 2012a).

A common strategy often recommended for overcoming the absorption problem involves the establishment of sovereign wealth funds, oil stabilization funds, and other similar solutions that are used by other resource-rich countries elsewhere to avoid the effects of volatility, fiscal excess, indebtedness, export-inhibiting exchange rates, and other problems (Naím 2009). In addition to controlling government spending, these policies are meant to ensure that revenues are saved and invested so that future generations may also benefit from the country's wealth even after it has been depleted. The main concern here is not over the effectiveness of these strategies but their blanket application in countries without regard to the national context, including the immediate developmental needs. For instance, some sovereign wealth fund programs, such as Botswana's Pula Fund, have been very effective in preventing the negative effects of excessive spending and price volatility. However, the effectiveness of this strategy in resource-rich countries with poor institutions is often limited because such funds either get raided before the rainy days or squandered in poor investments. The case of Libya's sovereign wealth fund, the Libyan Investment Authority, has been cited as an example of programs that became a victim of the Libyan government's inability to manage an ambitious strategy and the country's pervasive rent-seeking culture (Meijia and Castel 2012).

Perhaps a more serious problem with saving in external accounts to solve the so-called Dutch disease is the policy's possible stifling effects. As Collier and Warnholz (2009) have demonstrated, the rate of private return on investment in Africa is higher than in any other region. Yet Africa has not seen any significant influx of investment—foreign direct investment or otherwise (Asiedu 2004). Clearly, Africa is an investment-deficit region and with high potential returns, and therefore siphoning resources from the region and putting them in a low-returns savings market abroad amounts to stifling the potential for growth in the economy. Given the high rate of returns on investment in Africa, the solution by IMF and others that countries should "save abroad rather than invest domestically is costly defeatism" (Collier 2010: 133). Van der Ploeg (2012b) echoed this point and added that from a development perspective investing in sovereign wealth (e.g., US Treasury bills with a low rate of interest) rather than using the oil revenue to pay off foreign debt (which typically has a higher rate of interest) or invest in domestic infrastructure, education, or health projects (which typically have a much higher rate of return) does not make sense for many resource-rich African countries. Clearly, the problem for most African countries is capital scarcity and not overinvestment; and their main challenge is how to boost economic development, not how to engineer a sustained increase in consumption financed out of interest on US Treasury bills.

Investing in the continent requires detailed knowledge of the region and related capacities to choose good opportunities, but African countries can build or improve such capacities, learning from the expertise that has already been developed in this domain by others. Also, as

Collier (2010) argues, the real problem with domestic public investment programs in many developing countries is corruption rather than absorption. Without underestimating the potential for corruption in many countries, some researchers make a case for using the resource wealth to support public investment projects in public infrastructure and human development including investments in education, health, and nutrition (Van der Ploeg 2012b). A carefully designed program for managing natural resource revenues for broadly distributed human capital will not only help avoid the resource curse, but also make a positive difference for people living in resource-rich countries for many generations to come (Atinc 2013). Thus, a more targeted approach to the Dutch disease will be helpful.

In this context, Collier's (2010) three strategies for investing natural resource wealth in African countries, collectively called "investing to invest" for overcoming the absorption problem, deserve closer attention. These strategies include the need for the government to improve its management of public investment, the need for the government to induce private investment by improving the policy environment, and the need to contain the price of capital goods.

c) Alternative natural resource governance

Another anticipated change that will result from the good-fit approach involves governance arrangements that stress the role of civil society. As local communities have increasingly asserted their voice in natural resource exploration negotiations, especially in the mining sector, civil societies have emerged as important governance actors and have demanded a greater share of benefits for the long-term development of the local communities and increased their involvement in decision-making in natural resource management. The extractive industry developer's embrace of "social license to operate," local content strategies, and the desire to manage the relationship between local and large-scale miners are examples of how African governments can help chart out alternative natural resource governance that is more reflective of each nation's context.

Mineral developers around the world have embraced "social license to operate" (SLO) as a way of avoiding potentially costly conflicts and exposure to social risks (Prno and Scott Slocombe 2012). SLO initiatives were spurred by the growth of the sustainable development paradigm and governance shifts that have increasingly transferred governing authority toward non-state actors leading to pressures from local communities. According to the World Bank (2003), SLO involves acquiring "free, prior and informed consent of local communities and stakeholders." Pike (2012) adds that a full SLO must include both the acquisition and ongoing maintenance of the consent of the local stakeholders. The maintenance of consent is especially important as the criteria by which local stakeholders give their consent may change over time. Pike (2012) argues that better understanding of local conditions, equitability of the distribution of benefits from mining and land rights, artisanal versus illegal miners, the environmental impact of the project, and the track record of the company are especially important for maintaining SLOs. SLO is becoming increasingly relevant for mining companies due to their ability to ensure that local stakeholders remain supportive of projects.

Another strategy used by some oil- and gas-producing African countries to wade off local discontent is the introduction of requirements for "local content" into their regulatory frameworks (IPIECA 2011). Local content refers to strategies to ensure that outputs from the extractive industry sector generate further benefits to the local economy beyond the direct contribution of its added value, through its links to other sectors (Tordo and Anouti 2013). The requirements for local content strategy include creating jobs, promoting enterprise development, and accelerating the transfer of skills and technologies (IPIECA 2011). As Mwakali and Byaruhanga (2011) show, the experience of Nigeria and Angola with the implementation of local content strategy is riddled with contracts that were awarded to shell companies, cost inflation, increased project cycles, and the general slowdown in the oil and non-oil sectors. However, such experiences should not be used against local content strategy; rather, they should serve as reminders of how to minimize potential abuses of the strategy. Successful local content programs require good and clear understanding of the business benefits and a well-defined strategic plan, which is realistic and adapted to the country's situation. Consequently, detailed understanding of the local context is imperative for effective SLOs. Knowing the neighborhood also enables companies to understand the importance of local context in terms of requirements, capabilities, and the barriers that limit local worker and local company participation. These factors vary by location, the nature of the project, and the stage of the project life cycle (IPIECA 2011).

Finally, effective management of the relationships between large-scale mining companies and the local/indigenous artisanal and small-scale mining sector is critical for building domestic legitimacy for resource extraction and for reducing potential sources of conflicts over resource extraction. Managing the relationship must however start with a tacit recognition that small and artisanal mining has a legitimate and significant role to play in the social and economic development of these countries. It must also acknowledge their potential competition for the same mineralization, impacts on livelihoods if access to resources is limited, and changing social conditions, including conflict between artisanal small mining, host communities, and large-scale mining companies. Some representatives of large-scale mining operations have recently published a set of approaches and tools for mining companies to engage with artisanal small mining.[3] While this is a good beginning, governments must be encouraged to take this to the next level by addressing some of the demands of artisanal mining in their specific national contexts.

In sum, many

African countries have it within their grasp to build up governance structures [required for] the management of their resource wealth from short-term political pressures. [The key is ensuring that the institutions set up] for the purpose of immunizing from the vicissitudes of the political process... [are accompanied by requisite] fiscal, monetary, and exchange rate policies and institutions... to increase as far as possible the efficiency of revenue collection and to uproot the scourge of overvaluation.

(Gylfason 2011: 29)

Moving Africa beyond the resource curse

A decade of unprecedented growth, however uneven, coupled with the discovery and exploitation of valuable natural resources, has led many to anticipate the establishment of a set of "developmental states" in Africa to balance the many "fragile" or "failed" states that have dominated the continent's development narratives. This emergence of a "revised landscape" has coincided with a growing criticism of the claims of the resource curse theorists (Obi 2010), because the resource-conflict link is more sophisticated and complex than is often conceptualized in the literature (ACBF 2013). The notion of resource curse has also been criticized from a methodological and econometric perspective (Arthur 2012). Similarly, Humphreys et al. (2007) have called into question the resource curse argument, because for them there is considerable room for human agency to rectify the risks posed by the "paradox of plenty."

Such criticisms and challenges of the resource curse theory have led to a partial move away from the initial debates over the "greed versus grievance" causal binary. It is therefore not surprising that much of the emphasis on resource curse has shifted to issues related to capacity, leadership, and good governance (Obi 2010; ACBF 2013). We briefly highlight the centrality of these factors in the design and implementation of good-fit policies.

Capacity

The extant literature on capacity development is varied. According to ACBF (2011, 2013), capacity means having the aptitudes, resources, relationships, and facilitating conditions required to act effectively to achieve specified mandates. Capacity is conceptualized at three levels: individual, work environment or organizational, and institutional (interactions between individuals and organizations). Capacity also takes its meaning in specific settings (capacity for what?).

Linked to the natural resources sector in Africa, a number of institutional capacity challenges have been identified in the literature (IMF 2010). At the individual level, the capacity to forge a vision entails choices and leadership options—hence the need for transformational leadership capacity. Closely linked to leadership capacity is the need to provide training programs to both state and non-state actors (including media). As Ahonsi (2011) argues, countries that lack trained and skilled human resources essential for tackling critical challenges invariably struggle to resolve development issues. By advancing intellectual capital and human capacity, African countries could improve their individual and institutional performance.

At the institutional level, Africa faces myriad hurdles. Notable is the acute capacity deficit in enforcement mechanisms resulting from policy gaps, the pressures of globalization, insecurity, institutional decay, and inefficiencies linked to pervasive corruption (Chapter 4). Relatedly, there is a need to build capacity to ensure improved budgeting and expenditure management, improve procurement practices and grants of natural resources concessions, establish effective processes to control corruption, and support central institutions of government (ACBF 2013). The good-fit approach can help bridge these capacity gaps with sound policies, mindful of country-specific variables and contexts.

Clearly, the information and power asymmetry between African states and their external partners, stemming from their limited capacity, weaken African states' ability to engage in proper negotiations with foreign stakeholders in the natural resources sector. Barma et al. (2012) flag the issue of transfer pricing, which negatively impacts revenue collections in many resource-rich African states. While most African tax laws have legal provisions aimed at addressing the issue, these provisions are insufficient at best (Brautigam et al. 2008).

Again, given that Africa operates in a rapidly changing global environment, there is a need to embrace the knowledge economy to bridge the current knowledge gap. Consequently, African states must intensify investment in improving their knowledge infrastructure. Enhancing capacity to sustain effective auditing, monitoring, regulating, and improving resource exploitation regimes and developing resource sector linkages into the domestic economy should be another goal of African countries. This requires better physical infrastructure, and a dose of policies that are tailored to fit each specific country's context. Similarly, the capacity of the legislature in Africa to act as a countervailing force over the executive and undertake a strategic scanning of natural resource legislation is critical (Barma et al. 2012). Efforts should also aim at tackling the poor regulatory capacity of many resource-rich countries in Africa. African governments also need to urgently resolve resource beneficiation inequities (Obi 2010), which result in tensions and conflict, and impede development.

Going forward, African states should place emphasis on advancing capacity to actively engage in green markets, and strengthening the capacity of higher education and training institutions to provide requisite skills for all and to raise the skills base of the national workforce (ACBF 2013; see also Chapter 9).

Leadership

Given the rapidly changing natural resource management landscape in Africa, as highlighted in this volume, the centrality of leadership—transformative leadership—in navigating this dynamic landscape is critical (ACBF 2013; Chapters 2 and 8). Indeed, the pressure for change within the natural management sector has intensified with self-policing measures within the extractive industry (e.g., Extractive Industries Transparency Initiative (EITI); Kimberley Process Certification Scheme (KPCS); Publish What You Pay (PWYP)), heightened awareness and empowerment of CSOs, and growing calls for change. Adapting to the threats, opportunities, and possibilities requires a leadership that not only is transformative, but also has the unique ability to engage in strategic scanning, that is, the capacity to recognize the behavior of interconnected systems to make effective decisions under varying strategic and risk scenarios, and the transformation of knowledge: thus, a leadership that is politically astute, economically savvy, and business aware and uses its emotional intelligence to drive success (Hanson and

Léautier 2011), assess local political economy, and implement policies to improve natural resource management value chain.

As a strategic asset, leadership enhances capacity to (a) formulate policies and programs for development, (b) implement development initiatives, and (c) recognize the behavior of interconnected systems to make effective decisions under varying strategic and risk scenarios (Hanson and Léautier 2011). Contemporary leaders in the natural resources sector increasingly operate in very complex and interconnected environments. The degree of interconnectedness invariably shapes one's decision-making processes as well as the outcomes of their decisions. To this end, understanding the dynamics of the natural resources sector's interconnected environment is vital to (a) shaping strategy, (b) developing effective risk management approaches, and (c) selecting from a series of potential courses of action (Hanson and Léautier 2011).

Clearly, the transformational leadership required for implementing good-fit policies has to drive development policies that are most relevant to the country context and priorities: dialogue with stakeholders to invest funds from natural resources in the most relevant sectors (industry, infrastructure, knowledge, etc.); synergy with CSOs, giving them a voice and funding to ensure that they play their roles effectively; and guaranteeing equity in the redistribution of natural resource wealth or environmental/health concerns. Leadership should also ensure that transparency and accountability guide management decisions and serve as remedies for overcoming the curse (Humphreys et al. 2007). As UNDP (2011) also notes, the success stories of countries such as Norway, Botswana, Chile, Indonesia, and Malaysia are closely tied to transformational and development-oriented leadership, who were able to engage with strong constituencies outside the natural resources sector in the management of these countries' natural resources proceeds. By the same token, other resource-rich nations that have failed in this regard have met a rather sad fate (see Duruigbo 2006; Collier 2007; Humphreys et al. 2007).

Building effective transformational leadership capacity to design and implement good-fit policies is no easy task: the multiplicity of externalities, myriad stakeholders and interests, and a weak or nonexistent institutional environment, all impact on the process of advancing transformational leadership in Africa's natural resources sector (ACBF 2013). But, as Puplampu (Chapter 8) argues, while there is no one-size-fits-all model shaping the interaction between leadership, natural resources, and national development, inasmuch as resource wealth can propel the overall development of the society, citizens should demand accountability from their leaders (see also Humphreys et al. 2007).

Good governance

Scholars such as Collier (2007) note that abundant natural resources and the absence of good governance—accountability, transparency, the rule of law, and participation—often lead to greed and help turn a "resource into a curse" (see also Collier 2010; Arthur 2012). Within the context of Africa's divided societies, the governance of natural resources, particularly extractives, is very critical because control over natural resource beneficiation is often a prime motivator of ethnic or identity-based conflicts (ACBF 2013; see also Chapters 2, 3, and 4).

As Mehlum et al. (2006) note, resource mismanagement is likely to occur in countries that lack strong institutions and good governance. Similarly, Wantchekon (2002) contends that when state institutions are weak such that budget procedures either lack transparency or are discretionary, resource windfalls tend to help consolidate already-established government elite. It is as a result of such observations that many propose good governance as a panacea for resource mismanagement in Africa (ACBF 2013; Alao 2007). Governance for a good-fit approach requires as much transparency and accountability as possible and contributes to achieving overall development. To this end, having a media-literate citizenry and engaged civil society can help foster good enough governance for good-fit policies. Beyond the role and contribution of the media, effective resource management can be further enhanced if individuals and civil society have the political space for expression (Tettey 2013).

Another step to advance the efficient and effective governance of natural resources involves pursuing good-fit policies that ensure the equitable distribution of resource beneficiation. As Chikozho (2012), writing on transboundary natural resource management, argues, the higher the net beneficiation perceived, the higher the likelihood of cooperation is. To this end, working toward good governance necessitates an enabling environment and institutional structures that enhance stakeholder cooperation, but more importantly a strong determination to cooperate and coordinate efforts (Chikozho 2012). Indeed, a participatory governance process, at all levels, even if imperfect, can assist resource-rich countries attain sustainable economic growth and socioeconomic development (ACBF 2013; Tettey 2013).

Challenges do exist. As Barma et al. (2012) note, the sector is vulnerable to rent-seeking activities, and sector management vulnerabilities in the regulation and award of leases, in revenue collection and administration, and, eventually, in the way budget procedures secure sustainability in revenue reinvestment. This notwithstanding, there is no doubt that the net benefit of natural resources can only be improved with appropriate reforms in governance that are responsive to local conditions (Collier 2010; Barma et al. 2012; ACBF 2013).

Concluding words to the volume

As a way forward, this chapter joins the volume's authors to argue that African countries and stakeholders should enhance and advance efficient management and governance of natural resources by coordinating and integrating planning agencies operating across various sectors of the natural resources value chain. Equally important is the need to improve and maximize tax collection and utilize proceeds to ensure sustainable development. Other steps include, but are not limited to, enhancing leadership capacity, deepening capacity development interventions, advancing peer learning and knowledge sharing, and strengthening existing, while building accountable and transparent institutions, and adopting good-fit approaches and policies that can help African states manage the entire natural resource value chain—from exploration to negotiation, contracting, exploitation, and ultimately management and transformation of resource proceeds. A good case in point is the Africa Mining Vision. The formulation and articulation of the Africa Mining Vision would have been inconceivable before the current decade. Now the continent is better able to formulate its own responses to the post-2015 development agenda. At least some African actors can now claim a degree of agency rather than endure continuing dependency (Brown and Harman 2013): the prospect of policy-making rather than policy-taking. Africa inherited a ubiquitous North-South axis; it now needs to advance and benefit from an impending East-South turn.

Notes

1. For instance, Auty (2001) argues that "point resources" such as fuels and minerals engender policies that promote narrow sectional interests. However, "diffuse" resources like food and agricultural products are not significantly correlated with institutional quality.
2. The NRM value chain spans the key sequence of steps that a resource-dependent country must undertake in transforming its natural resources into developmental aspirations. These include sector organization and contract awards, regulation and monitoring operations, collection of taxes and royalties, revenue distribution and management, and development of sound and sustainable policies (ACBF 2013).

3. See a report titled *Working Together: How Large-Scale Mining Can Engage with Artisanal and Small-Scale Miners*, and published in 2010 by The International Council on Mining and Metals (ICMM), the International Finance Corporation's Oil, Gas and Mining Sustainable Community Development Fund (IFC CommDev), and Communities and Small-Scale Mining (CASM), available at: http://www.icmm.com/page/17638/new-publication-on-engaging-with-artisanal-and-small-scale-miners. Accessed on September 15, 2013.

References

ACBF. (2011). *African Capacity Indicators Report: Capacity Development in Fragile States*, Harare, Zimbabwe: African Capacity Building Foundation.

ACBF. (2013). *Africa Capacity Indicators Report: Capacity Development for Natural Resources Management*, Harare, Zimbabwe: African Capacity Building Foundation.

Ahonsi, B. (2011). "Capacity and Governance Deficits in the Response to the Niger Delta Crisis," in C. Obi and S. Rustad (eds.), *Oil and Insurgency in the Niger Delta: Managing the Complex Politics of Petroviolence*, London: Zed Books, pp. 28–41.

Alao, A. (2007). *Natural Resources and Conflict in Africa: The Tragedy of Endowment*, Rochester, NY: University of Rochester Press.

Arthur, P. (2012). "Averting the Resource Curse in Ghana: Assessing the Options," in L. Swatuk and M. Schnurr (eds.), *Natural Resources and Social Conflict: Towards Critical Environmental Security*, London: Palgrave Macmillan, pp. 108–127.

Asiedu, E. (2004). "Policy Reform and Foreign Direct Investment to Africa: Absolute Progress But Relative Decline," *Development Policy Review*, 22(1): 41–48.

Atinc, T. (2013). "Avoiding the Resource Curse: How to Manage Natural Resource Wealth for Human Development," *Up Front*, Brookings Institution, July 11, Available at: http://www.brookings.edu/blogs/up-front/posts/2013/07/11-natural-resource-wealth-africa-atinc. Accessed September 7, 2013.

Auty, R.M. (2001). *Resource Abundance and Economic Development*, Oxford: Oxford University Press.

Barma, N.H., Kaiser, K., Le, T.M., and Viñuela, L. (2012). *Rents to Riches? The Political Economy of Natural Resource-Led Development*, Washington, DC: World Bank.

Besada, H., (2013). "Doing Business in Fragile States: The Private Sector, Natural Resources and Conflict in Africa," Background Research Paper, Submitted to the High Level Panel on the Post-2015 Development Agenda (May).

Brautigam, D., Fjeldstad, O.-H., and Moore, M. (eds.) (2008). *Taxation and State-Building in Developing Countries: Capacity and Consent*, Cambridge: Cambridge University Press.

Brown, W. and Harman, S. (eds.) (2013). *African Agency in International Politics*, Abingdon, UK: Routledge.

Bulte, E.H., Damania, R., and Deacon, R.T. (2005). "Resource Intensity, Institutions, and Development," *World Development*, 33(7): 1029–1044.

Chikozho, C. (2012). "Towards Best-Practice in Transboundary Water Governance in Africa: Exploring the Policy and Institutional Dimensions of Conflict and Cooperation over Water," in K. Hanson, G. Kararach, and T.M. Shaw (eds.), *Rethinking Development Challenges for Public Policy*, London: Palgrave Macmillan, pp. 155–200.

Collier, P. (2007). *The Bottom Billion: Why the Poorest Countries Are Failing and What Can Be Done about It*, New York: Oxford University Press.

Collier P. (2010). *The Plundered Planet: Why We Must—And How We Can—Manage Nature for Global Prosperity*, New York: Oxford University Press.

Collier P. and Warnholz, J.L. (2009). "Now's Is the Time to Invest in Africa," *Harvard Business Review*, Breakthrough Ideas for 2009, Available at: http://hbr.org/web/2009/hbr-list/now-the-time-to-invest-in-africa. Assessed September 15, 2013.

Corden, W. and Neary, J.P. (1982). "Booming Sector and Deindustrialization in a Small Open Economy," *Economic Journal*, 92(368): 825–848.

Diamond, L. and Mosbacher, J. (2013). "Petroleum to the People: Africa's Coming Resource Curse—And How to Avoid It," *Foreign Affairs*, September/ October, Available at: http://www.foreignaffairs.com/articles/139647/larry-diamond-and-jack-mosbacher/petroleum-to-the-people. Accessed September 7, 2013.

Duruigbo, E. (2006). "Permanent Sovereignty and Peoples' Ownership of Natural Resources in International Law," *International Law Review*, 33: 44–62.

Grindle, M. (2004). "Good Enough Governance: Poverty Reduction and Reform in Developing Countries," *Governance: An International Journal of Policy, Administration and Institutions*, 17: 525–548.

Grindle, M. (2007). "Good Enough Governance Revisited," *Development Policy Review*, 25(5): 533–574.

Gylfason, T. (2004). *Natural Resources and Economic Growth: From Dependence to Diversification*, London: CEPR (Center for Economic Policy Research Discussion Paper 4804).

Gylfason, T. (2011). *Natural Resource Endowments: A Mixed Blessing?* Munich, Germany: CESifo (CESifo Working Paper 3353).

Hanson, K. and Léautier, F.A. (2011). "Enhancing Institutional Leadership in African Universities: Lessons from ACBF's Interventions," *World Journal of Entrepreneurship, Management and Sustainable Development*, 7(2–4): 386–417.

Humphreys, M., Sachs, J.D., and Stiglitz, J.E. (2007). *Escaping the Resource Curse*, New York: Columbia University Press.

IMF (International Monetary Fund). (2010). "IMF Launches Trust Fund to Help Countries Manage Their Natural Resource Wealth," Press Release No. 10/497, December 16, 2010, Available at: http://www.imf.org/external/np/sec/pr/2010/pr10497.htm. Accessed September 26, 2013.

IPIECA. (2011). *Local Content Strategy: A Guidance Document for the Oil and Gas Industry*, London: IPIECA.

Mehlum, H., Moene, K.O., and Torvik, R. (2006). "Institutions and the Resource Curse," *Economic Journal*, 116(5): 1–20.

Meijia, P.X. and Castel, V. (2012). "Could Oil Shine like Diamonds? How Botswana Avoided the Resource Curse and its Implications for a New Libya," *AfDB Chief Economist Complex*, Tunis, Tunisia: The African Development Bank, October.

Mwakali, J.A. and Byaruhanga, J.N.M. (2011). "Local Content in the Oil and Gas Industry: Implications for Uganda," in J.A. Mwkali and H.M. Alinaitwe (eds.), *Proceedings of the 2nd International Conference on Advances in Engineering and Technology*, Entebbe, Uganda, January 31–February 1, 2011, Kampala: Macmillan Uganda (Publishers) Ltd., pp. 517–522.

Naím, M. (2009). "The Devil's Excrement: Can Oil-Rich Countries Avoid the Resource Curse?" *Foreign Policy*, September/October, Available at: http://www.foreignpolicy.com/articles/2009/08/17/the_devil_s_excrement. Accessed September 7, 2013.

Obi, C. (2010). "Oil Extraction, Dispossession, Resistance, and Conflict in Nigeria's Oil-Rich Niger Delta," *Canadian Journal of Development Studies*, 30(1–2): 219–236.

Oshewolo, S. and Oniemola, R.M. (2011). "The Financing Gap, Civil Society, and Service Delivery in Nigeria," *Journal of Sustainable Development in Africa*, 13(2): 254–268.

Pike, R. (2012). "The Relevance of Social License to Operate for Mining Companies," Schroders Social License to Operate Research Paper, July.

Prno, J. and Scott Slocombe, D. (2012). "Exploring the Origins of 'Social License to Operate' in the Mining Sector: Perspectives from Governance and Sustainability Theories," *Resources Policy*, 37: 346–367, Available at: http://dx.doi.org/10.1016/j.resourpol.2012.04.002. Accessed September 7, 2013

Sachs, J.D. and Warner, A.M. (1995). "Natural Resource Abundance and Economic Growth," NBER Working Paper 5398, Cambridge, MA: National Bureau of Economic Research (NBER).

Sachs, J.D. and Warner, A.M. (2001). "The Curse of Natural Resources," *European Economic Review*, 45: 827–838.

Strauss, M. (2000). "The Growth and Natural Resource Endowment Paradox: Empirics, Causes and the Case of Kazakhstan," *Fletcher Journal of Development Studies*, 16: 1–28.

Tettey, W.J. (2013). *Media and Information Literacy, Informed Citizenship and Democratic Development in Africa: A Handbook for Media/Information Producers and Users*, Harare: ACBF.

Tordo, S. and Anouti, Y. (2013). *Local Content in the Oil and Gas Sector: Case Studies*. Washington, DC: World Bank.

UNDP. (2011). *Managing Natural Resources for Human Development in Low-Income Countries*, Washington, DC: UNDP—Regional Bureau for Africa.

Van der Ploeg F. (2012a). "Bottlenecks in Ramping Up Public Investment," *International Tax and Public Finance*, 19(4): 509–538.

Van der Ploeg, F. (2012b). "Managing Oil Windfalls in the CEMAC," in B. Aikitoby and S. Coorey (eds.), *Oil in Central Africa—Policies for Inclusive Growth*, Washington, DC: International Monetary Fund, pp. 89–110

Wantchekon, L. (2002). "Why Do Resource Abundant Countries Have Authoritarian Governments?" Available from: http://www.afea-jad.com/2002/Wantchekon3.pdf. Accessed September 26, 2013.

World Bank. (2003). *Striking a Better Balance: The World Bank and Extractive Industries*, Washington, DC: World Bank.

Afterword

Natural Resource Governance Post-2015: What Implications for Analysis and Policy?

Timothy M. Shaw

> The South has risen at an unprecedented speed & scale... By 2050, Brazil, China & India combined are projected to account for 40 percent of world output in purchasing power parity terms...
>
> The changing global political economy is creating unprecedented challenges and opportunities for continued progress in human development.
>
> (UNDP 2013: 1–2)

The above opening citation from the 2013 UNDP *Human Development Report* on the rise of the Global South brings together global governance, natural resources, human development, and security. This concluding reflection on the state of natural resource governance (NRG) is informed by animating a new PhD at the University of Massachusetts Boston on Global Governance and Human Security, and continuing after three decades to edit an *International Political Economy* (IPE) Series with a focus on the Global South, in which this collection is located. It has also been informed by the mid-2013 North South Institute Forum on "Governing Natural Resources for Africa's Development" and two special workshops at the annual International Studies Association conferences in early-2013 and 2014 on the political economy of energy. I build on the increasingly familiar and compatible concepts of "the transnational" (Hale and Held 2012) and "global governance" (Harman and Williams 2013; Weiss and Wilkinson 2014a, b) as together they advance the analysis of natural resources in Africa symbolized by the Kimberley Process (KP) and the Extractive Industries Transparency Initiative (EITI).

The first decade of the 21st century was that of the BRICs/BRICS, especially China and India, leading Jan Nederveen Pieterse (2011: 22) to assert that the established North–South axis is being superseded by an East–South one:

> ...the rise of emerging societies is a major turn in globalization...North–South relations have been dominant for 200 years and now an East–South turn is taking shape. The 2008 economic crisis is part of a global rebalancing process.

An earlier, longer version of this chapter appeared as the Foreword to the Classics pb reprints published in the IPE Series in the fall of 2013, usually pp. ix–xxii.

The post-2015 global political economy

To situate NRG post-2015, this conclusion/projection juxtaposes a set of parallel and overlapping perspectives to consider whether the several "worlds"—from North Atlantic/Pacific and onto Eurozone of PIIGS (Portugal, Italy, Greece, and Spain) versus "second world" (Khanna 2009) of BRICS/CIVETS/MINT/MIST/VISTA[1]—have grown together or apart as global crises and reordering have proceeded (see myriad heterogeneous analyses such as Cooper and Antkiewicz 2008; Cooper and Subacchi 2010; O'Neill 2011; Pieterse 2011; Economist 2012; Lee 2012; USNIC 2012; WEF 2012; World Bank 2012; Cooper and Flemes 2013; Gray and Murphy 2013).

In turn, "contemporary" "global" issues—wide varieties of ecology, gender, governance, health, norms, technology, etc. (see section *"Emerging 'global' issues"* below)—have confronted established analytic assumptions/traditions and actors/policies leading to myriad "transnational" coalitions and heterogeneous initiatives, processes, and regulation schemes as overviewed in Bernstein and Cashore (2008), Dingwerth (2008), and Hale and Held (2011) (see section "Informal and illegal economies: From fragile to developmental states?"); these impact prospects for NRG in Africa as elsewhere. Such extra or semi-state hybrid "global governance" increasingly challenges and supersedes exclusively interstate international organization and law (Harman and Williams 2013; Weiss and Wilkinson 2014a, b).

Each set of Emerging Markets (EMs) embody slightly different sets of assumptions, directions and implications; Pricewater Coopers (PWC) expanded the "Next-11" of Goldman Sachs (i.e., 15 without South Africa) to 17 significant EMs by 2050 (Hawksworth and Cookson 2008): what implications for NRG? Symptomatically, the initial iconic acronym was proposed at the start of the new century by a leading economist working for a global financial corporation—Jim O'Neill (2011) of Goldman Sachs (www2.goldmansachs.com)—who marked and reinforced his initial coup with celebration of its first decade. As he notes, global restructuring has been accelerated by the simultaneous decline not only of the United States and United Kingdom but also of the southern members of the Eurozone. Many now predict China to become the largest economy by 2025 and India to catch-up with the US by 2050 (Hawksworth and Cookson 2008: 3). PWC (2013: 6, 8) suggests that

> The E7 countries could overtake the G7 as early as 2017 in PPP terms...the E7 countries could potentially be around 75% larger than the G7 countries by the end of 2050 in PPP terms...

> By 2050, China, the US and India are likely to be the three largest economies in the world....

But Stuart Brown (2013: 168–170) notes that there are competing prophecies about the cross-over date when China trumps the United States, starting with the International Monetary Fund (IMF) advancing it to 2016.

As the G8 morphed into G20 (Cooper and Antkiwicz 2008; Cooper and Subbachi 2010), a variety of analysts attempted to map the emerging world, including Parag Khanna's (2009) "second world' and Fareed Zakaria's (2011) "rest." For example:

a) the World Economic Forums' Global Redesign Initiative (GRI) which included a small state caucus centered on Qatar, Singapore, and Switzerland (Cooper and Momani 2011);[2]
b) the Constructive Powers Initiative advanced by Mexico (www.consejo mexicano.org/en/constructive-powers) which brought older and newer middle powers together (Jordaan 2003) such as the old Anglo Commonwealth with inter alia Indonesia, Japan, and South Korea;
c) the US National Intelligence Council, at the end of 2012, from both sides of the pond, produced "Global Trends 2030: alternative worlds" (GT 2030) (www.gt2030.com) which identified four "megatrends" like "diffusion of power" and "food, water, energy nexus"; a half-dozen "game-changers"; and four "potential worlds" from more to less conflict/inequality, including the possibilities of either China–US collaboration or of a "non-state world"; and
d) Chatham House in London reported on "Resources Futures" (Lee 2012: 2) with a focus on "the new political economy of resources" and the possibility of NRG by "Resource 30" (R30) of major producers/consumers, importers/exporters (www.chathamhouse.org/resourcesfutures): G20 including the BRICs, but not BRICS (i.e., no South Africa), plus Chile, Iran, Malaysia, Netherlands, Nigeria, Norway, Singapore, Switzerland, Thailand, United Arab Emirates, and Venezuela.

And in the case of the most marginal continent, Africa, its possible renaissance was anticipated at the turn of the decade by the Boston Consulting Group, the Center for Global Development, McKinsey et al. (Shaw 2012a), with the *Economist* admitting in January 2011 that it might have to treat Africa as the "hopeful" rather than "hopeless" continent. Meanwhile, the supply of development resources is also moving away from the old North toward the BRICS (Chin and Quadir 2012) and other new official donors like South Korea and Turkey plus private foundations like Gates, faith-based organizations, remittances from diasporas, sovereign wealth funds, and novel sources of finance such as taxes on carbon, climate change, emissions, and financial transactions (Besada and Kindornay 2013).

Emerging economies/states/societies?

The salience of "emerging markets" especially the BRICS and other political economies in the second world has led to debates about the similarities and differences among emerging economies/middle classes/multinational companies/states/societies, etc. Informed by different disciplinary cannons, for example, in contrast to Goldstein (2007) on emerging economies multinational companies (EMNCs), Pieterse (2011) privileges sociologically informed emerging societies. In turn, especially in international relations, there are burgeoning analyses of emerging powers, regional and otherwise (Jordaan 2003; Flemes 2010; Nel and Nolte 2010; Nel 2012), some of which might inform new regionalist perspectives, especially as these are increasingly impacted by the divergence between

BRICS and PIIGS. In turn, they inform and advance alternative definitions of, and directions for, development.

Despite the US subprime and EU euro-crises at the start of the 21st century, foreign direct investment (FDI) in Africa continues to grow, reaching US$50 billion in 2013: primarily from China, India, and Malaysia. With new energy discoveries and investments, a second tier of oil producers has emerged after Nigeria and Angola: Equatorial Guinea, Congo-Brazzaville, Gabon, South Sudan, and now Ghana, with Uganda eager to join. Liquefied natural gas (LNG) is now exported from Nigeria, Equatorial Guinea, and Mozambique, with the latter projected to challenge the dominance of Qatar and Australia by 2020? And by mid-2014, the World Bank plans to launch a US$ one billion fund to map the continent's mineral resources to advance the Africa Mining Vision.

Varieties of development

"Development" was a notion related to post-war decolonization and bipolarity. It was popularized in the "Third World" in the 1960s, often in relation to "state socialism," one party even one man rule, but superseded by neo-liberalism and the Washington Consensus. Yet the newly industrialized countries (NICs), then BRICs, pointed to another way by contrast to those in decline like fragile states (Brock, Holm, Sorensen, and Stohl 2012); such "developmentalism" (Kyung-Sup, Fine, and Weiss 2012) has now even reached Africa (UNECA 2011, 2012). But, while the "global" middle class grows in the South, so do inequalities along with non-communicable diseases (NCDs) like cancer, coronary heart disease, and diabetes. Given the elusiveness as well as limitations of the Millennium Development Goals (Wilkinson and Hulme 2012), the UN has been debating and anticipating post-2015 development desiderata (www.un.org/millenniumgoals/beyond2015) including appropriate, innovative forms of governance as encouraged by networks around international nongovernmental organizations (www.beyond2015.org) and think tanks (www.post2015.org)[3] Aid is now about cooperation and not finance as a range of flows is attracted to the global south including private capital, foreign direct investments, philanthropy/faith-based organizations (FBOs), remittances, let alone money laundering (Shaxson 2012); official development assistance (ODA) is a shrinking proportion of transnational transfers (Brown 2011, 2013: 24–28). Meanwhile, around such dramatic global reordering, the varieties of capitalisms, state and non-state proliferate.

Varieties of capitalisms

The world of capitalisms has never been more diverse: from old trans-Atlantic and trans-Pacific to new – the global South with its own diversities such as Brazilian, Chinese, Indian and South African "varieties of capitalisms"? Andrea Goldstein (2007) introduced emerging economies multinational companies in the *International Political Economy Series*, including a distinctive second index: five pages of company names of EMNCs (see next paragraph). And in the post-neo-liberal era, state-owned enterprises (SOEs), especially national oil companies (Xu 2012), are burgeoning. Both US/UK neo-liberal, continental and Scandinavian corporatist, as well as Japanese and East Asian developmentalist "paradigms," are having to rethink and reflect changing state-economy/society relations beyond ubiquitous "partnerships" (Overbeek and van Apeldoorn 2011). Furthermore, if we go beyond the formal and legal, then myriad informal sectors and transnational organized crime and money laundering are ubiquitous.

For the first time, in the "Global Fortune 500" of (July) 2012, multinational companies' headquarters were more numerous in Asia than in either Europe or North America. There were 73 Chinese multinational companies so ranked (up from 11 a decade ago in 2002) along with 13 in South Korea and eight each in Brazil and India. Each of the BRICS or EMs hosted some global brands: Geely, Huawei, and Lenovo (China), Hyundai, Kia, and Samsung (Korea); Embraer and Vale (Brazil); Infosys, Reliance, and Tata (India); and Anglo American, De Beers, and SABMiller (South Africa).

The pair of dominant economies in sub-Saharan Africa is unquestionably Nigeria and South Africa yet, despite being increasingly connected, they display strikingly different forms of "African" capitalisms and NRG. Nigeria, including its mega-cities like Lagos and Ibadan, is a highly informal political economy with a small formal sector (notably beer, consumer goods such as soft drinks and soaps, finance, and telecommunications); by contrast, despite its ubiquitous shanty-towns, South Africa is based on a well-established formal economy centered on mining, manufacturing, farming, finance, and services. Both have significant diasporas in the global North, especially the United Kingdom and United States, including Nigerians in South Africa, especially Johannesburg, remitting funds back home. Since majority democratic rule, South African companies and supply chains, brands, and franchises have penetrated the continent; initially into Eastern from Southern Africa, but now increasingly into West Africa and Angola.

New regionalisms

The proliferation of states along with capitalisms post-bipolarity has led to a parallel proliferation of regions, especially if diversities of non-state, informal even illegal regions are so considered. And the Eurozone crisis concentrated in the PIIGS has eroded the salience of the EU as model leading to a recognition of a variety of "new" regionalisms (Flemes 2010; Shaw et al. 2011). These include instances of "African agency" (Lorenz and Rempe 2013) like South African franchises and supply chains reaching to West Africa and the Trilateral Free Trade Area among the Common Market for East and Southern Africa (COMESA), the East Africa Community (EAC), and the Southern African Development Community (SADC) (T-FTA) (Hartzenberg 2012) along with older/newer regional conflicts such as the Great Lakes Region (GLR) plus the regional as well as global dimensions of, say, piracy off the coast of Somalia (ACBF 2014; Hanson 2014).

Emerging "global" issues

A growing number of global issues are increasingly recognized arising in the global South as well as resulting from excessive consumption and pollution in the North such as NCDs such as diabetes. In the immediate future, these issues will include environmental and other consequences of climate change and health viruses/zoonoses. They will also extend to myriad computer viruses and cyber-crime (Kshetri 2013). Some suggest that we may be running out of basic commodities such as energy (Klare 2012) and water, let alone rare-earth elements (REEs). Finally, after recent global and regional crisis, the governance of the global economy is at stake: the financialization syndrome of DBRAs (Debt Bond Rating Agencies), derivatives, exchange traded funds/exchange traded notes, hedge and pension funds, and sovereign wealth funds (Overbeek and Apeldoorn 2012).

Informal and illegal economies: From fragile to developmental states?

Developing out of the Internet, new mobile technologies increasingly facilitate the informal, illegal as well as otherwise. The "informal sector" is increasingly recognized in the discipline of anthropology, etc., as the illegal in the field of international political economy (Naylor 2005; Friman 2009); these are increasingly informed by telling Small Arms Survey (SAS) annual reports after more than decade with a focus on fragile states (www.smallarmssurvey.org).

Similarly, organized crime is increasingly transnational with the proliferation of (young and male) gangs from myriad states (see Knight and Keating 2010: chapter 12). In response, the field of international political economy needs to develop analyses and prescriptions from the established informed annual Small Arms Survey (SAS) to the Global Commission on Drug Policy and Health (www.globalcommissionondrugs.org) which builds on the successful experience of the Latin American Commission on Drugs and Democracy. And in 2012, Google Ideas (www.google.com/ideas/events/info-2012/) set up the Illicit Networks to appreciate the challenges of illicit arms trade, money laundering, etc. Further, as supply chains shifted from Central America and the Caribbean to West Africa in response to the 'war on drugs', the Kofi Annan Foundation created a preventive West African Commission on Drugs (www.wacommissionondrugs.org).

Varieties of transnational governance

Just as "governance" is being redefined and rearticulated (Bevir 2011), so the "transnational" is being rediscovered and rehabilitated (Dingwerth 2008; Hale and Held 2012) following marginalization after its initial articulation at the start of the 1970s by Keohane and Nye (1972): they identified major varieties of transnational relations such as communications, conflict, education, environment, labor, multinational companies, and religions. And Stuart Brown (2011) updated such perspectives with a more economics-centered framework which included civil society and remittances.

In turn, I would add contemporary transnational issues such as brands and franchises; conspicuous consumption by emerging middle classes; world sports, such as Fédération Internationale de Football Association (FIFA) and the International Olympic Committee (IOC); global events from World Fairs to Olympics and world soccer; logistics and supply chains (legal, formal, and otherwise); mobile digital technologies; new film centers such as Bollywood and Nollywood including diasporas, film festivals, and tie-ins; new media such as Facebook and Twitter. But such heterogeneous relations and or perspectives including the KP, the EITI, and the African Mining Vision (AMV) deserve further attention in terms of their contribution to NRG in Africa and elsewhere.

Global governance and natural resources by mid-century?

In conclusion, I juxtapose a trio of changes which will probably impact global governance and natural resources in policy and practice in Africa and elsewhere post-2015:

i) exponential global restructuring in myriad areas, from economics and ecology to diplomacy and security (Overbeek and Apeldoorn 2012; Besada and Kindornay 2013);

ii) shift in the direction and concentration of supply chains away from South–North toward South–East; and

iii) continued evolution in multi-stakeholder communities to incorporate state-owned enterprises, sovereign wealth funds, pension funds, and exchange traded funds, especially from the BRICS and other EMs.

Notes

1. Columbia, Indonesia, Vietnam, Egypt, Turkey, and South Africa (CIVETS); Mexico, Indonesia, Nigeria, and Turkey (MINT); Mexico, Indonesia, South Korea, and Turkey (MIST); and Vietnam, Indonesia, South Africa, Turkey and Argentina (VISTA).
2. For a readers' guide to GRI, see: www.umb.edu/cgs/research/global_redesign_ initiative
3. See the section on "Varieties of transnational governance."

Bibliography

ACBF. (2014). *Africa Capacity Indicators Report 2014: Building Capacity for Regional Integration in Africa,* Harare: The African Capacity Building Foundation.

Bernstein, Steven and Cashore, Benjamin (2008). "The Two-Level Logic of Non-State Market Driven Global Governance," in Volker Rittberger and Martin Nettesheim (eds.), *Authority in the Global Political Economy,* London: Palgrave Macmillan, pp. 276–313.

Besada, Hany and Kindornay, Shannon (eds.) (2013). *The Future of Multilateral Development Cooperation in a Changing Global Order,* London: Palgrave Macmillan for NSI.

Bevir, Mark (2011). *Sage Handbook of Governance,* London: Sage.

Brock, Lothar, Holm, Hans-Henrik, Sorensen, Georg and Stohl, Michael (2012). *Fragile States,* Cambridge: Polity.

Brown, Stuart (2013). *The Future of US Global Power: Delusions of Decline,* London: Palgrave Macmillan.

Brown, Stuart (ed.) (2011). *Transnational Transfers and Global Development,* London: Palgrave Macmillan.

Chin, Gregory and Quadir, Fahim (eds.) (2012). "Rising States, Rising Donors and the Global Aid Regime," *Cambridge Review of International Affairs*, 25(4): 493–649.

Cooper, Andrew, F. and Antkiewicz, Agata (eds.) (2008). *Emerging Powers in Global Governance: Lessons from the Heiligendamm Process*, Waterloo: WLU Press for CIGI.

Cooper, Andrew, F. and Momani, Bessma (2011). "Qatar and Expanded Contours of South-South Diplomacy," *International Spectator*, 46(3): 113–128.

Cooper, Andrew, F. and Flemes, Daniel (eds.) (2013). "Special Issue: Emerging Powers in Global Governance," *Third World Quarterly*, 34(6): 943–1144.

Cooper, Andrew, F. and Shaw, Timothy, M. (eds.) (2013). *Diplomacies of Small States: Resilience Versus Vulnerability?* London: Palgrave Macmillan. Second edition.

Cooper, Andrew, F. and Subacchi, Paola (eds.) (2010). "Global Economic Governance in Transition," *International Affairs*, 86(3): 607–757.

Cornelissen, Scarlett, Fantu Cheru and Shaw, Timothy, M. (eds.) (2012). *Africa and International Relations in the Twenty-first Century: Still Challenging Theory?* London: Palgrave Macmillan.

Dingwerth, Klaus (2008). "Private Transnational Governance and the Developing World," *International Studies Quarterly*, 52(3): 607–634.

Dunn, Kevin, C. and Shaw, Timothy, M. (eds.) (2013). *Africa's Challenge to International Relations Theory*, London: Palgrave. Revised Classics pb edition.

Economist (2012). *The World in 2013*, London www.economist.org/theworldin/ 2013.

Fanta, Emmanuel, Shaw, Timothy, M. and Tang, Vanessa (eds.) (2013). *Comparative Regionalism for Development in the Twenty-first Century: Insights from the Global South*, Farnham: Ashgate for NETRIS.

Fioramonti, Lorenzo (eds.) (2012). *Regions and Crises: New Challenges for Contemporary Regionalisms*, London: Palgrave Macmillan.

Flemes, Daniel (ed.) (2010). *Regional Leadership in the Global System*, Farnham: Ashgate.

Friman, H. Richard (ed.) (2009). *Crime and the Global Political Economy*, Boulder: LRP. IPE Yearbook #16.

Goldstein, Andrea (2007). *Multinational Companies from Emerging Economies*, London: Palgrave Macmillan.

Gray, Kevin and Murphy, Craig (eds.) (2013). "Special Issue: Rising Powers and the Future of Global Governance," *Third World Quarterly*, 32(2): 183–355.

Hale, Thomas and David Held (eds.) (2012). *Handbook of Transnational Governance*, Cambridge: Polity.

Hanson, Kobena, T. (ed.) (2014). *Contemporary Regional Development in Africa*, Farnham Ashgate.

Hanson, Kobena, T., Kararach, George and Shaw, Timothy, M. (eds.) (2012). *Rethinking Development Challenges for Public Policy: Insights from Contemporary Africa*, London: Palgrave Macmillan for ACBF.

Hartzenberg, Trudi et al. (2012). *The Trilateral Free Trade Area: Towards a New African Integration Paradigm?* Stellenbosch: Tralac.

Hawksworth, John and Cookson, Gordon (2008). "The World in 2050: Beyond the BRICs: A Broader Look at Emerging Market Growth," London: PWC.

Jordaan, Eduard (2003). "The Concept of Middle Power in IR: Distinguishing Between Emerging and Traditional Middle Powers," *Politikon*, 30(2): 165–181.

Jordaan, Eduard (2012). "South Africa, Multilateralism and the Global Politics of Development," *European Journal of Development Research*, 24(2): 283–299.

Keohane, Robert, O. and Nye, Joseph, S. (eds.) (1972). *Transnational Relations and World Politics*, Cambridge: Harvard University Press

Khanna, Parag (2009). *The Second World: How Emerging Powers are Redefining Global Competition in the Twenty-First Century*, NY: Random House.

Klare, Michael, T. (2012). *The Race for What's Left: The Global Scramble for the World's Last Resources*, New York: Metropolitan.

Kliman, Daniel, M. and Fontaine, Richard (2012). "Global Swing States: Brazil, India, Indonesia, Turkey and the Future of International Order," Washington, DC: GMF.

Knight, W. Andy and Keating, Tom (2010). *Global Politics*, Toronto: OUP Chapter 12.

Kshetri, Nir (2013). *Cybercrime and Cybersecurity in the Global South*, London: Palgrave Macmillan.

Kugelman, Michael (2009). *Land Grab? Race for the World's Farmland*, DC: Brookings

Kyung-Sup, Chang, Fine, Ben and Weiss, Linda (eds.) (2012). *Developmental Politics in Transition: The Neoliberal Era and Beyond*, London: Palgrave Macmillan.

Lee, Bernice et al. (2012). *Resources Futures*, London: Chatham House, December www.chathamhouse.org/resourcesfutures.

Lorenz-Carl, Ulrike and Rempe, Martin (eds.) (2013). *Mapping Agency: Comparing Regionalisms in Africa*, Farnham: Ashgate.

Margulis, Matias, E. et al. (eds.) (2013). "Special Issue: Land Grabbing and Global Governance," *Globalizations*, 10(1).

Naylor, R. T. (2005). *Wages of Crime: Black Markets, Illegal Finance and the Underworld Economy*, Ithaca: Cornell University Press. Second edition.

Nel, Philip et al. (eds.) (2012). "Special Issue: Regional Powers and Global Redistribution," *Global Society*, 26(3): 279–405.

Nel, Philip and Detlef Nolte (eds.) (2010). "Regional Powers in a Changing Global Order," *Review of International Studies*, 36(4): 877–974.

O'Bryne, Darren, J. and Hensby, Alexander (2011). *Theorising Global Studies*, London: Palgrave Macmillan.

OECD (2012). *Economic Outlook, Analysis and Forecasts: Looking to 2060. Long-Term Growth Prospects for the World*, Paris.

O'Neill, Jim (2011). *The Growth Map: Economic Opportunity in the BRICs and Beyond*, New York: Portfolio/Penguin.

Overbeek, Henk and Apeldoorn, Bastiaan van (eds.) (2012), *Neoliberalism in Crisis*, London: Palgrave Macmillan.

Pieterse, Jan Nederveen (2011). "Global Rebalancing: Crisis and the East-South Turn," *Development and Change*, 42(1): 22–48.

Power, Marcus, Mohan, Giles and Tan-Mullins, May (2012). *China's Resource Diplomacy in Africa: Powering Development*, London: Palgrave Macmillan.

PWC. (2013). *World in 2050: The BRICs and Beyond: Prospects, Challenges and Opportunities*, London, January.

Ratha, Dilip et al. (2011). *Leveraging Remittances for Africa: Remittances, Skills and Investments*, Washington, DC: World Bank and AfDB.

Reuter, Peter (ed.) (2012). *Draining Development: Controlling Flows of Illicit Funds From Developing Countries*, Washington, DC: World Bank. www.openknowledge.worldbank.org.

Shaw, Timothy, M. (2012a). "Africa's Quest for Developmental States: 'Renaissance' for Whom?" *Third World Quarterly*, 33(5): 837–851.

Shaw, Timothy, M., Cooper, Andrew, F. and Chin, Gregory, T. (2009). "Emerging Powers and Africa: Implications for/from Global Governance?" *Politikon*, 36(1): 27–44

Shaw, Timothy, M., Grant, J. Andrew and Cornelissen, Scarlett (eds.) (2011). *Ashgate Research Companion to Regionalisms*, Farnham: Ashgate.

Shaxson, Nicholas (2012). *Treasure Islands: Uncovering the Damage of Offshore Banking and Tax Havens*, New York: Palgrave Macmillan.

SID. (2012). *State of East Africa 2012: Deepening Integration, Intensifying Challenges*, Nairobi for TMEA.

Sinclair, Timothy (2012). *Global Governance*, Cambridge: Polity.

Singh, Priti and Izaralli, Raymond (eds.) (2012). *The Contemporary Caribbean: Issues and challenges*, New Delhi: Shipra.

Sumner, Andy and Mallett, Richard (2012). *The Future of Foreign Aid*, London: Palgrave Macmillan.

UNDP. (2012a). *African Human Development Report: Towards a Food Secure Future*, New York, May.

UNDP. (2013). *Human Development Report 2013: The Rise of the South: Human Progress in a Diverse World*, New York, March.

UNDP. (2014). *Human Development Report 2014: Beyond 2015: Accelerating Human Progress and Defining Goals*, New York, March.

UNECA. (2011). *Economic Report on Africa 2011: Governing Development in Africa: The Role of the State in Economic Transformation*, Addis Ababa.

UNECA. (2012). *Economic Report on Africa 2012: Unleashing Africa's Potential as a Pole of Global Growth*, Addis Ababa.

USNIC. (2012). *Global Trends 2030: Alternative Worlds*, DC: National Intelligence Council, December, www.dni.gov/nic/globaltrends www.gt.com

Vom Hau, Matthias, Scott, James and Hulme, David (2012). "Beyond the BRICs: Alternative Strategies of Influence in the Global Politics of Development," *European Journal of Development Research*, 24(2): 187–204.

Weiss, Thomas, G. (2013). *Global Governance*, Cambridge: Polity.

Weiss, Thomas, G. and Wilkinson, Rorden (eds.) (2014a). *International Organization & Global Governance*, Abingdon: Routledge.

Weiss, Thomas, G. and Wilkinson, Rorden (2014b). "Global Governance to the Rescue: Saving International Relations?" *Global Governance*, 20(1): 19–36.

Wilkinson, Rorden and Hulme, David (eds.) (2012). *The Millennium Development Goals and Beyond: Global Development After 2015*, Abingdon: Routledge.

World Bank. (2012). *Global Economic Prospects June 2012: Managing Growth in a Volatile World*, DC, June.

World Economic Forum. (2012). *Global Risks 2012: An Initiative of the RRN*, Davos. Seventh edition.

Xing, Li with Farah, Abdulkadir Osman (eds.) (2013). *China-Africa Relations in an Era of Great Transformation*, Farnham: Ashgate.

Xu, Yi-chong (ed.) (2012). *The Political Economy of State-Owned Enterprises in China and India,* London: Palgrave Macmillan.

Xu, Yi-chong and Baghat, Gawdat (eds.) (2011). *The Political Economy of Sovereign Wealth Funds,* London: Palgrave Macmillan.

Zakaria, Fareed (2011). *The Post-American World: Release 2.0,* New York: Norton. Updated and expanded edition March 2014.

Index

Note: The letters 'f', 'n' following locators refer to figures and notes respectively.

Aborigines' Rights Protection
    Society, 145
Abuja declaration, 201
Addis Ababa Declaration on
    Sustainable Biofuels
    Development, 194
Africa
  bunkering, 83
  capacity building by universities, 29
  capacity development, 54–5
  carbon tax, 198
  civil society organizations, 29
  coal mining, 187
  commodity demand, 96
  contributor to global climate, 186
  cotton export (Mali), 101–2
  criminality in natural resource
    chain in, 83–7
  diamond trade (Botswana), 102–3
  export share, 93
  FDI, 229
  fuel price protests, 196
  GHG emissions contribution, 187–8
  global financial crisis, 196
  governance-related issues, 43
  green growth, 191–4, 200
  gross domestic product (GDP), 4
  growth period, 92
  import, 97
  infrastructure, 191
  investment-deficit region, 214
  knowledge sharing, 7
  leadership, 175–7
  management initiatives, 44–9
  market-friendly economy, 18
  mis-management of natural
    resources, 67
  multinational corporations, 18–20
  NRM governance, 23–5
  oil demand in, 94
  oil production (Angola), 103–4
  policy choices, 175–7
  political leadership, 166–72
  poor resource management, 27
  postcolonial African state, 17–18
  post-conflict transition, 69
  recommendations for good NRM,
    30–2, 55–60
  renaissance in NRM, 20–3
  solar energy, 194
  state corporations, 18–20
  state role in NRM, 165
  tax collection, 21
  transboundary river basins, 120f
  urbanization in, 94, 96
  water scarcity, 120
African
  Africa Carbon Support Programme
    (ACSP), 196
  Africa Mining Vision (AMV), 5,
    222, 229
  African Capacity Building
    Foundation (ACBF), 2, 156, 200
  African Development Bank (AfDB),
    6, 92, 94, 156, 185–6, 198–9
  African Development Bank's African
    Green Fund (AfGF), 198
  African Minerals Development
    Centre (AMDC), 6
  African Mining Vision (AMV),
    5–6, 232
  African Peer Review Mechanism
    (APRM), 6, 175; critical
    review, 175
  African Union Commission,
    54–5, 57–9
  Africa Progress Report, 4
agricultural commodities, 9, 16, 92,
    97, 113
Algeria
  active oil producing country, 166
  commodity export, 93

Algeria – *continued*
  natural gas export, 97–8
  oil price rises, 99–100
  OPEC affiliation, 166
American cotton subsidies, 97
Amoco, 142
Anadarko Petroleum Corporation, 143
Angola
  commodity export, 93
  crude oil production, 103
  economic growth, 103
  exploitation of extractive natural
    resources, 39
  financial resource allocations, 50
  national energy security
    priorities, 171
  new energy discoveries, 229
  oil price rises, 99–100
  oil sales revenue, 104
  oil wealth, 167
  OPEC membership, 167
  overseas scholarships, 50
  self-financing conflict, 42
Angolan Sonangol, 142
armed insurgencies, 67, 70
Atlantic Resources, 71–2
Australian Petroleum Production &
  Exploration Association, 151

best practices, 10, 24, 32, 52, 88, 207,
  209, 211
biofuel crop production, 193–4, 201
Botswana
  Citizen Entrepreneurial
    Development Agency, 112
  development-oriented
    leadership, 31
  diamond industry, 40, 101–3,
    111–12, 168
  economic dependence on
    diamond, 102
  good governance, 60
  joining of EITI, 49
  judicial system, 58
  natural resources management, 32,
    57–8, 60
  political leadership in, 173
  pre-modern growth, 102
  property rights, 58

Pula Fund, 198, 214
rainy-day fund, 56
unequal income distributions, 103
where good governance, 58
BRICS model, 2
budgeting and expenditure
  management, 7, 73, 218
bureaucracy
  ideal-type features, 164
  postcolonial bureaucracy, 174
  state bureaucracy, 173

Cameroon
  forestry in, 21
  management of natural
    resources, 48
  oil rent, 50, 172
  state affairs, 50
capacity development, 54–5, 119–20
  bureaucratic capacity, 26
  challenges, 25–7
  civil society organizations'
    advocacy, 29
  deepening interventions, 32
  definition, 54, 119
  governments concessions, 26
  International Monetary Fund
    initiatives, 28
  learning experiences of various
    countries, 29–30
  legislature, 26–7
  opportunities, 27–30
  pricing transfer, 26
  regulatory capacity, 26
  Revenue Watch Institute
    initiatives, 28
  royalties, 26–7
  systematic approach, 10
  tax concessions, 26–7
  traditional approaches, 119
  United Nations University-Institute
    for Natural Resources in Africa
    (UNU-INRA), 28–9
  World Bank initiatives, 27–8
Capacity Development for the
  Environment (CDE), 157
Capacity Development Strategic
  Framework (CDSF), 54, 57,
  60, 61n4

capacity-leadership-governance trichotomy, 3
capitalisms varities, 229–30
Cap-Net global network, 191
carbon limits, 197–9
carbon tax, 197–9
cartelization, 94
cash crop, 97, 99
Certified Emission Reduction units (CERs), 196–9
Chad
  cotton dependency, 97
  development-oriented leadership, 31, 220
  escrow account, 175
  Future Generations Fund, 175
  non-OPEC country, 166
  oil discovery, 174–5
  oil revenue, 176
  public sector's ineffectiveness, 55
ChevronTexaco, 171
Chile, natural resources management, 32
civil society
  capacities to inform communities, 154–5
  capacities to propose mitigation, 155
  capacity of, 53
  connections with transnational advocacy networks, 52
  democratic participation of, 52
  environmental governance, 154–8
  in Ghana, 145–47
  importance of, 51–4
  information sharing for risk management, 156–7
  integration of, 83
  in mining, 151–4
  monitoring capacities, 155
  negotiation capacities, 155–6
  in oil sector, 145, 147–51
  partnership and funding, 158
  policy-making, 156
  proliferation of, 6
  regional cooperation, 156–8
  risk cycle, 157f
  role of, 52, 168, 212, 215
  society groups, 3, 8, 52–3, 72, 83, 86–7, 175, 215
  strategic role, 53
  transnational investment, 52
  vibrancy of, 175
  watchdog, creation of, 175
Civil Society Coordinating Council (CivisoC), 146
Clean Development Mechanism (CDM), 196–7
Clean Technology Fund (CTF), 199
Climate Development Knowledge Network (CDKN), 190–1
Climate Investment Funds' (CIFs), 199
cogeneration, 195, 202
colonialism, 106, 118
colonial legacy, 16–17
commodity
  agricultural, 92, 113
  boom-bust cycle, 96
  earning, 98
  export, 15, 91, 93–4, 112
  index, 96f
  long-term trends of price, 96
  non-renewable, 41
  price fluctuation, 92
  price volatility, 98
  super-cycle, 96
Common African Defence and Security Policy (CADSP), 87
comparative advantage theory, 106
conflict minerals, 11
Congo
  exploitation of natural resources, 39
Constructive Powers Initiative, 228
Convention on the Conservation of Nature and Natural Resources, 21
corporate income taxes, 21
corporate social responsibility (CSR), 20–1
corruption
  civil society's role, 212
  external dimensions of, 177
  high incidence, 50
  inefficiencies linked to, 67
  institutionalization of, 41
  mechanisms to curb, 7, 73–4
  resource curse, 107

criminality, 3, 9, 66–9, 75–6, 80, 82–7
  armed conflicts, 85
  colluding government officials, 84
  colluding state security agencies, 85
  criminal syndicates profiteering, 86
  cross-border and regional
    dimensions, 85, 87
  illegal logging, 69–71
  illicit timber trade, 68–9
  impacts of, 85
  international criminal networks, 85
  key dynamic, 67
  Liberia's timber industry, 69
  local communities and their
    inhabitants, 85
  logging syndicate, 71–2
  natural resource value chain, 86–7
  in Nigeria's oil value chain, 75
  in oil sector, 67
  private use permits, 68–75
  reform agenda, 73
  revenue loss, 85
  sanction regime to PUPs regime,
    72–5
  scale of, 86
  timber resource management, 67
  visibility of, 86
criminal syndicates profiteering,
  84, 86
CTF funds, 199
current commodity-driven export, 93
cyclical commodity boom-and-bust
  cycle, 96

Danube Basin Commission, 125
democratic governance, decline
  of, 163
dependency theory, 105–6, 108
development
  biofuel, 193–4
  capacity development; challenges,
    25–7; in NRM, 54–5;
    opportunities, 27–30
  clean development mechanism,
    196–7
  consensus-building approach, 122
  development-oriented
    leadership, 31

economic, 17, 20, 41, 43, 97, 123,
  129, 208
  environmental, 187
  framework, 163
  green technology, 189
  infrastructural, 42
  long-term strategy, 101
  national strategies, 187–8
  socialist path, 17
  socioeconomic, 40, 44–5, 50, 55, 57,
    128, 130, 170, 175
  statist approach, 17
  structuralist theories, 105
  sustainable, 172–5, 186, 191, 210–1
  theory, 21
  varieties, 229
  water resource, 45, 124
diagnostic-based risk-adjustment
  system, 231
diamond
  beneficiation program, 111
  blood diamonds, 46
  consumption, 167
  entrepreneurialism programs, 112
  export, 102, 111
  Kimberley Process Certification
    Scheme, 175
  lack of transparency, 47
  marketing, 167
  on-shoring of, 111
  processing, 46
  smuggling, 47
  trading, 46
  value-added processing of, 111
dispensation, 4–7
  community-based NRM, 5
  consumer spending, 4
  corporate social responsibility, 4
  missing revenues, 4
  returns on investments, 4
Dodd-Frank bill, 11
Dutch disease, 21, 23, 32, 60, 104,
  106–8, 145, 163, 172, 208, 213–15

East Africa Community (EAC), 230
Economic Community of West
  African States (ECOWAS), 83

economic development
  in basin, 123, 132
  commodity production, 97
  ecological sustainability and, 128
  important factor for, 43
  through clean energy, 185
economic diversification, 5, 9, 59, 92,
    100, 104–11, 112–13
  dependency theory, 105–6
  Dutch disease, 106–8
  import substitution, 108–9
  modernization theory, 104–5
  new structural economics,
    109–10
  resource curse theories, 106–8
  trust fund approach, 110–11
  world systems theory, 105–6
economic importance, 25
economic recession, 91–2
Economic Report on Africa, 4
economic responsibility, 3, 141
The Economics of Welfare, 197
effective utilization, 162
Egypt
  natural gas export, 97–8
  renewable energy initiatives, 194,
    199, 202
Egyptian hydro-hegemony, 45
emerging markets, 228–9
environment
  business-friendly, 18
  degradation, 42, 45, 101, 124
  governance, 3, 140–59; capacities
    required for, 154–8; civil society
    organizations and, 151–4;
    defined, 141; effectiveness for
    oil sector in Ghana, 141;
    framework for, 151; linkage
    with development, 141;
    mining and, 151–4; principles
    of, 152–3
  environmental impact assessment
    (EIA), 149
  investor-friendly, 20
  long-term impact, 141
  normative, 6
  relation with economic
    development, 173

Rio+20 declarations, 173, 176, 186,
    199, 201
environmental impact assessment
    (EIA), 149
Environmental Protection Act, 152
environmental sustainability, 3, 10,
    66, 71, 85, 88, 141, 162, 195
  challenges for, 195
  MDGs attainment, 162
  principles of, 3, 66
Equatorial Guinea
  deplorable living conditions, 171
  national energy security
    priorities, 171
  new energy discoveries, 229
  non-OPEC country, 166
  oil-producing countries, 167
  resource nationalism, 169
ethanol conversion, 193, 201
Ethiopia
  capacity building, 190
  fertilizers credit scheme, 192
  railway transport system, 188
  renewable energy strategies, 199
Eurozone, 227, 230
expertise, 16, 49, 58–9, 82, 147, 156,
    186, 214
exploitation of natural resources
  criminal, 84
  illegal, 46, 85
  MNCs role, 16
  political structure, 16
  postcolonial state, role of, 8
  social and environmental cost, 21
  sustainable form of, 56
  unsustainable, 122
export trends, 93–8
external and internal vulnerability,
    98–100
extractive industries
  government influence on, 24
  inadequate flow of net benefits, 24
  private enrichment over public
    welfare, 211
  revenue diversions from, 47
  transparency in financial
    transactions, 47
  transparency initiatives, 85
  value chains in, 9, 113

Extractive Industries Transparency
Initiative (EITI), 2, 5, 8–9, 11, 23,
40, 47–9, 53, 60, 87, 176, 197,
219, 226, 232
ExxonMobil, 171

failed state, 61n1
fairness, 3, 66
farming, subsistence-based, 110
financing mechanisms, 195–6
flying geese paradigm, 110, 113
Food and Agriculture Organization
(FAO), 68
foreign direct investment (FDI), 73–4,
189, 214, 229
Forestry Development Authority
(FDA), 69–70
formal-legal authority, 164
Friends of the Nation (FoN), 152
Future Generations Fund, 175
The Future We Want, 176

Gabon
monitoring and management of
government agencies, 48
new energy discoveries, 229
non-OPEC country, 166
oil-producing countries, 167
Ghana
civil society groups' role, 52–3
land grabbing debate, 193
mining sectors, 24
new energy discoveries, 229
oil or mining legislation, 4
oil sectors, 24
socialist path to development, 17
Ghana National Petroleum
Corporation (GNPC), 142
Global Commission on Drug Policy
and Health, 231
Global Environmental Facility
(GEF), 125
Global Fortune 500, 230
global governance, 2, 177, 226–7, 232
globalization, 19, 67, 84, 87, 163–5,
170, 218, 226
aspect of, 165
categories, 165
global skeptics, 165

hyperglobalists, 165
policy-making, 165
Global Redesign Initiative (GRI), 228
Global West Vessel Specialist Agency
(GWVSA), 82
Global Witness, 70–1, 74
good enough governance, 209–10, 221
good-fit approach, 7, 206–22
governance for, 221
leading proponents, 209
prerequisites for, 207, 211–17
good-fit policies, 10, 207, 209–11, 218,
220–1
good governance, 7–8, 17, 30, 39–40,
42, 49–50, 53, 57–8, 60, 147, 209,
218, 221
accountability, 49–55
capacity development, 54–5
civil society, 51–4
informed media, 51–4
media's role and contribution, 51
participation, 49–55
pillars of, 51
rule of law, 49–55
transparency, 49–55
Governance and Economic
Management Assistance Program
(GEMAP), 72
achievements in forestry
management, 73
aim of, 73
FDA's capacity, 73
objectives, 73
governance of natural resources, 15,
60, 221–2
Governance Partnership Facility
(GPF), 53
governance-related problems, 43
greed versus grievance, 2, 44, 218
greener/clean energy
development, 185
green growth, 3, 10, 185–202
agricultural transformation, 187
agro-ecological condition, 188
basic tenets of, 187
biofuel crop production, 194
capacity building in human capital
and skills, 189–91
carbon limits, 197–9

carbon tax, 197–9
certified emission reduction units, 197–9
clean development mechanism (CDM), 196–7
climate aid to emission reductions, 186
comparative advantage, 187–8
concentrated solar power (CSP), 194
cost benefits, 187
drought-resistant crops, 191
environmental risk/benefit assessment, 188
feasibility of, 191–4; agricultural intensification, 192; biofuels, 192–4; chemical fertilizers, 191–2; critical elements of sustainable development, 192; green economy, 192; green revolution movement, 192
fertilizer subsidy scheme, 188
food security, 187
genetically modified crops, 189
green growth, 187
greenhouse gas emissions contributors, 187
growth development strategy, 188
infrastructure, 191
intellectual property rights, 189
intergenerational implications, 186
labor-intensive renewable energy sector, 189
long-term goal, 202
mismatch with development strategies, 200–1
national development strategies, 187–8
new green jobs, 189
political economy, 187–8
population growth, 195
promotion of, 6
regional and global solutions, 199–200
regulation, 195
shortage of green engineers, 189
technology implications, 188–9
Green Growth Trust Fund, 200
greenhouse gas (GHG) emissions, 196
green revolution, 187, 191–2, 201

gross domestic product (GDP), *see under individual countries*
guiding principle of policy-makers, 145

heterogeneous initiatives, 227
human development index, 169, 172
human rights violations, 171
hydraulic fracturing, *see* fracking
HYDRONIGER, 122
hydropolitics, 128, 131

identity-based conflicts, 15
identity-based political calculations, 176
inequitable distribution of benefits, 56
informal and illegal economies, 227, 231
infrastructural underdevelopment, 42
integrated economic system, 16
Integrated Social Development Center (ISODEC), 149
integrated water resources management (IWRM), 125, 127
internal and external pressures, 5, 133
International Council on Mining and Metals (ICMM), 156
International Law Commission, 129
International Maritime Organization, 79, 149
International Monetary Fund (IMF), 25, 28, 206, 213–14, 218, 227
International Olympic Committee (IOC), 231
International Resource Panel (IRP), 5
International Union for Conservation of Nature (IUCN), 131
investing to invest, 215

Jubilee Field, 142–4, 149

Kenya
    Civil Society Coalition on Oil and Gas, 147
    climate change mitigation, 192
    commodity export, 93
    food security, 192
    green growth, 185

Kenya – *continued*
  land grabbing debate, 193
  renewable energy strategies, 199
Kimberley Process, 2, 5, 11, 23, 40, 46,
  81, 175, 219, 226
Kimberley Process Certification
  Scheme (KPCS), 5, 8, 23, 40, 46–7,
  60, 175, 219
Kosmos Energy, 143

Lake Chad Basin Commission (LCBC),
  123–4, 129
Latin American Commission on Drugs
  and Democracy, 231
Laurelton Diamonds, 111
leadership
  broad categories, 163
  bureaucratic, 164
  critical perspectives on, 164
  deficit, 175
  definition of, 164
  development-oriented, 31
  essentialist perspective, 163
  form of, 164
  governance crisis and, 164
  managerial, 164
  multiple meanings of, 4
  nature of, 164
  non-existence of definite
    account, 164
  policy choices, 175–7; structural
    transformation, 176
  relational perspectives, 164
  sustainability and, 3
  transformational, 31
legal and negotiation skills
  shortage, 24
Liberia, 68–75
  business and diplomacy model, 70
  civil war, 70–2
  conflict logging, 69, 71
  cycle of violence, 71
  economic development, 69
  economic growth, 69
  economic stagnation, 74
  exploitation of natural resources, 39
  foreign direct investment
    (FDI), 73–4
  illegal logging, 69–71

inequitable distribution of
  benefits, 56
  logging syndicate, 71–2
  NGO Coalition for, 72
  oil or mining legislation, 4
  political stability, 69
  private use permits (PUP), 71
  sanction regime to PUPs
    regime, 72–5
  timber industry criminality, 69
Libreville Declaration on Biodiversity
  and Poverty Alleviation, 199
Libreville Declaration on Health and
  Environment, 199
Libya
  active oil producing country, 166
  natural gas export, 97–8
  OPEC affiliation, 166
life expectancy, 11
The Limits to Growth, 172

macro-economic stability, 18
Mali, 101–2
  COMATEX (state-run textile
    industry), 102
  cotton dependency, 97
  gold trade, 101
  long-term development
    strategy, 101
  over-farming of cotton, 101
  political troubles, 101
  position of Cotton in political
    economy, 101
  renewable energy strategies, 199
  structural adjustment program, 101
management initiatives, 44–9
  Extractive Industries Transparency
    Initiative (EITI), 47–9
  Kimberley Process Certification
    Scheme (KPCS), 46–7
  multilateral cooperative
    approach, 46
  regional power trade, 45
  resource management, 45
  transboundary environmental
    action, 45
  transboundary management, 44–6
Marikana crisis, 22
Maritime Security Agency, 82

market-friendly economies, 17
Mekong Water Dialogues, 131
Millennium Development Goals
    (MDGs), 162, 192, 229
mining
    law and regulation, 22
    Marikana crisis, 22
    nationalization of, 22
    transformation of, 22
modernization theory, 104–5
    multiple critiques of, 105
    practical implications of, 104
mono-industry economy, 41
Morocco
    energy conservation in
        transport, 199
    green growth, 185
    renewable energy initiatives, 194,
        199, 202
Mozambique
    biofuel, 187
    dependency on one
        commodity, 100
    discovery of oil and gas sources,
        11, 98
    drought-resistant crops, 190
    green growth, 185
    liquefied natural gas (LNG)
        export, 229
multinational corporations
    adversarial relations with
        community, 32
    concessions by governments, 26
    confidentiality clauses by, 24
    CSR initiatives by, 21
    dominance of, 18–19
    power balance with government, 19
    proliferation of, 20
    resources exploitation, 13, 20
    unfavourable contracts with
        government, 25

Namibia, 21, 45, 124
National Forestry Reform Law (NFRL),
    71, 73
National Hydrocarbons Company, 49
National Patriotic Front of Liberia
    (NPFL), 70
national sovereignty, 30, 128

Natural Resource, 5
    Natural Resource Charter (NRC), 5
    Natural Resources and
        Environmental Governance
        (NREG) program, 27
    regulation, 31–2
    taxation, 21
neoliberalism, critics of, 109
New and Renewable Energy
    Authority, 194
New Partnership for Africa's
    Development (NEPAD's), 58
new public management, 17
new regionalisms, 230
new structural economics, 109–10
Niger Basin Authority (NBA), 122
Nigeria
    absence of strong institutions, 50
    accountability, 50
    commodity export, 93
    criminality issues in oil sector, 67
    energy conservation in
        transport, 199
    exploitation of extractive natural
        resources, 39
    foreign exchange earnings, 50
    illegal bunkering in the oil sector,
        75–87
    inequitable distribution of
        benefits, 56
    insurgency, 41
    land grabbing debate, 193
    liquefied natural gas (LNG)
        export, 229
    militant activities, 41
    mining sectors, 24
    national energy security
        priorities, 171
    natural gas export, 97–8
    new energy discoveries, 229
    oil export, 27
    oil price rises, 99–100
    oil sectors, 24
    OPEC affiliation, 166
    political transparency, 50
    self-financing conflict, 42
    state of affairs in, 50
    violent conflict, 41, 58

Nigerian Extractive Industries
    Transparency Initiative
    (NEITI), 81
Nile Basin Initiative (NBI), 40, 45–6
    principles of, 45
Nile River Basin Commission (NRBC),
    45–6
Nolte, Detlef, 228
Non-OPEC African, 166
non-state actors
    ability of, 40
    capacity development programs, 58
    emergence of, 163
    relation with states, 163

Official Development Assistance
    (ODA), 197
oil bunkering, 75–87
    cargo theft format, 77
    collection of kickbacks for
        contracts, 77
    foreign actors, 80; international
        black marketers, 80; vessels
        crew/seafarers, 80
    illegal syndicate, 79–80
    local actors, 79–80; armed groups,
        79; colluding security
        personnel, 79–80; criminal
        gangs, 79; local artisans, 80;
        local communities, 79; local
        entrepreneurs, 80
    mushrooming of local refineries, 78
    nature of, 76–8
    official illegal bunkering, 76–7
    pipeline vandalization, 77
    policy responses, 80–3; capacity
        building, 81–2; concessioning
        maritime operations, 82;
        enhanced security, 80;
        international proposal on blood
        oil, 81; reducing incentives to
        trade, 81
    scale of, 78–9
    Togo Triangle, 77
    unofficial illegal bunkering, 77
    wellhead tapping, 77
oil certification (fingerprinting)
    scheme, 81

Oil sector of Ghana, 142–5
    challenge management, 149
    civil society action, 154–8; capacity
        imbalance, 158; CSOs role, 159;
        funding capacities, 158;
        mitigation measures, 155;
        monitoring capacities, 155;
        negotiation capacities, 155–6;
        partnership, 158; prevention,
        156; regional cooperation, 156;
        risk management, 156
    civil society organizations, 145–7,
        151–4
    Civil Society Platform, 148
    corporate social responsibility, 153
    economic consequences of oil
        exploitation, 145
    environmental and social impact
        assessment (ESIA), 154
    environmental and social
        impacts, 152
    environmental concerns, 149
    environmental governance, 151–4
    environmental impact assessment,
        149–50
    environmental legislation
        framework, 152
    Jubilee Field, 143–4, 149–50, 154
    legal frameworks, 149
    local fishermen livelihoods, 150
    marginal fishing exploitation, 150
    mining, 151–4
    Petroleum Exploration and
        Production Bill, 148
    Petroleum Revenue Bill, 148
    regulatory framework, 152
oil spills, 42, 149–50
oil stabilization funds, 206, 213
OKACOM, 124
Okavango River basin, 124–5
OPEC African countries, 167
openness, 3, 66
Orange River basin, 45, 61n3
Orange-Senqu River Commission
    (ORASECOM), 45
Organisation for Economic
    Co-Operation and Development
    (OECD), 119

Organization of Petroleum Exporting
  Countries (OPEC), 94, 166
Oxfam, 52–3, 97, 147, 149, 152,
  155, 206

paradox of plenty, 1, 7, 16, 32, 41, 66,
  162, 217
The Paradox of Plenty, 162
Petro-Canada, 142
Petroleum Host Community
  Development Fund, 82
Pigovian tax, 197
point resources, 222n1
policy recommendations, 30–2, 55–60
  capacity constraints, 58
  capacity development
    improvement, 59
  cooperation among riparians, 56
  cost- and benefit-sharing
    agreements, 56
  free public education, 57
  high-quality game parks, 57
  policy harmonization, 56
  resource management training, 59
  stakeholder cooperation, 56
  transboundary water
    cooperation, 55
  wildlife management, 57
political leadership and strategic
  choices, 166–72
  analytical framework, 168
  capacity deficiencies, 169
  diamond management, 168
  of oil wealth in national
    development, 166
  politicization of oil markets, 169
  resource nationalism, 169
postcolonial African state, 17–18
  affectation, 17
  anti-development, 17
  overdeveloped, 17
  precapitalist, 17
  predatory, 17
  prismatic, 17
  rolling back the state, 17
  soft, 17
  vampire, 17
  weak, 17

post-2015 global political economy,
  227–8
poverty
  alleviation program, 162
  environmental degradation, 101
  green growth, role of, 186
  growth, 43
  levels of, 11, 25
  material deprivation and, 84–5
  reduction expenditure, 175
  regional disparity and, 140
Pricewater Coopers (PWC), 227
private corporations, nationalization
  of, 18
private use permits (PUPs), 73
Programme for the Endorsement of
  Forest Certification (PEFC), 2, 9
Protocol on Shared Watercourses, 40
Publish What You Pay (PWYP), 5, 8,
  23, 86–7, 149, 219

rare-earth elements (REEs), 231
Reducing Emissions from
  Deforestation and Forest
  Degradation (REDD+), 6, 196, 198
Regional Economic Communities,
  121, 129
regional economic integration, 110,
  113, 121, 126, 134
  enhancement of, 126
  flying geese paradigm, 110, 113
  objectives of, 134
regulatory regimes, inadequacies
  of, 74
renaissance in natural resource
  management, 20–3
renewable natural resources, revenue
  from, 21
rent-seeking activities, 24, 208, 221
resource
  depletion, 41
  exploitation regimes, 7, 59, 219
  extraction, 5, 9, 21, 24, 41, 48, 54,
    92, 107, 110, 113, 158, 207, 216
  management value chain, 8, 211
  nationalism, 20, 32–3
  sector linkages, 7, 59, 219

resource-based economies
  agricultural exports, 92–3
  cyclical commodity export
    boom, 92
  dominance of commodities, 93
  foreign involvement in, 98
  land grabs, 93
  price rise, 92
resource curse
  best-fit policies, 211
  capacity, 218–19; individual level,
    218; institutional level, 218;
    organizational level, 218
  challenges, 44
  civil society, 210
  criticism, 43–4, 217
  dangers of, 9
  econometric and measurement
    fallacies, 43
  emergence of a revised
    landscape, 217
  explanation of, 208
  good-fit approach implementation,
    211–17; alternative resource
    governance, 215–17; Dutch
    disease, 213–15; stakeholders
    role, 212–13
  good governance, 221–2
  institutional explanations, 208
  leadership, 219–21; centrality of
    leadership, 219; governance for
    a good-fit approach, 221;
    resource beneficiation, 221;
    transformational leadership,
    220–1
  natural resource value chain, 210–11
  natural thesis, 107
  non-inevitable, 207–9
  one-size-fits-all solutions, 211
  political economy approach, 209–11
  poor regulatory capacity, 219
  rent-seeking models, 208
  slower growth, 208
  state failure, 42
  theory, 8, 44, 106–8, 218
  trade-based proxies, 43
responsible community development,
  3, 141
Revenue Watch, 28, 52, 206

Rio+20, 173, 176, 186, 199, 201
risk cycle, 157*f*

SADC Protocol on Shared
  Watercourses, 45
Scaling Up Renewable Energy
  Program, 198–9
scarce commodities, 20
Sierra Leone
  capacity of civil society, 53
  exploitation of natural resources, 39
  self-financing conflict, 42
Small Arms Survey (SAS), 231
social contract, 9, 30
socialist path to development, 17
social license to operate (SLO), 215–16
Society of Petroleum Engineers, 151
Southern African Development
  Community (SADC), 40, 129, 230
Sovacool, B., 39, 41, 43
sovereign authority, 2
sovereign wealth funds (SWFs), 4, 11,
  173, 206, 213–14, 228, 231–2
state capitalism, 18
state involvement in natural resource
  management, 18
state-owned enterprises (SOEs), 18
state-society relations, 165, 175
Strategic Climate Fund (SCF), 199
structural adjustment period, 97
structural adjustment programs
  (SAPs), 17, 101, 146, 186
structural transformation, 3, 5, 176
Sudan
  new energy discoveries, 229
  non-OPEC country, 166
  oil-producing countries, 167
sustainable budget index, 30
sustainable development, 172–5
  development debate, 173
  political independence, 174
  postcolonial bureaucracy, 174
Sustainable Energy Fund for Africa
  (SEFA), 198

Tanzania
  climate change mitigation, 192
  discovery of oil and gas sources, 98
  food security, 192

land grabbing debate, 193
socialist path to development, 17
taxation policies, 105, 177
tax reform, 197
tax revenue, paucity of, 25
Third World debt crisis, 109
Tiffany & Co., 111
timber value chain, 69–71
Tobga Timber Company, 71
Togo Triangle, 77
trade theory, 200
Train the Trainer programs, 191
Transboundary Diagnostic Analysis (TDA), 134
transboundary resource management, 3, 40, 45
transboundary river basin management (TRBM), 118
transboundary river basin organizations (TRBOs), 44
transboundary water resources, 10, 118, 129–30
    approaches and imperatives, 127–33
    basin communities, 131–2
    capacity building, 119–20, 126, 134
    civic participation, 131
    consensus-building approach, 122
    diagnostic analysis framework, 132
    domestic water management institutions, 130
    ecosystem goods and services, 132
    funding, 133
    governance framework, 130
    hot spots, 118, 125
    human capacity building, 127
    hydropolitics, 131
    ineffectiveness of leadership, 128
    institutional capacity building, 127
    international cooperation arrangements, 121
    international experiences, 125–6
    key objective, 119
    Lake Chad Basin Commission, 123–4
    legal frameworks, 128
    Mekong and Danube River basins, 125
    multi-stakeholder water dialogues, 130

national water management institutions, 129
    neutral dialogue platforms, 130–1
    Niger Basin Authority, 122
    Okavango River basin, 124–5
    organizational capacity building, 127
    policies and institutions, 129–30
    power dynamics, 130–1
    public-private partnerships, 133
    rigid allocation of organizational responsibilities, 130
    river basins in Africa, 120*f*
    scientific information generation, 132–3
    shared watercourse institutions, 129
    stakeholders, 131–2
    state-based approaches, 132
    tensions reduction among riparian, 127
    water conflict, 121
    water resources management, 128
    water sharing, 120–1
    water-wars thesis, 121
    win-win proposition, 119
transformation, axis of, 3
transitional groups, 100
transnational coalitions, 227
transnational governance, 9–10, 231–2
transnational organized crime, 230–1
transparency, principles of, 3, 66
tree crops, 99
trust fund approach, 110–11
trust funds/SWFs development, 6
Tullow Group Scholarship Scheme, 59–60
Tullow Oil, 59, 143
2013 Africa Capacity Indicators Report, 4
2014 Resource Governance Index, 4

Uganda
    Civil Society Coalition for Oil, 147
    discovery of oil and gas sources, 98
    new energy discoveries, 229
unemployment, 11–12, 42
UNESCO-Equatorial Guinea International Prize, 171

United Nations
  United Nations Conference on
    Environment and Development
    (UNCED), 201
  United Nations Educational,
    Scientific and Cultural
    Organization (UNESCO), 171
  United Nations Environment
    Programme (UNEP), 5
  United Nations Framework
    Convention on Climate Change
    (UNFCCC), 198
  UN Security Council, 72
urbanization patterns, 95*f*

value chain
  capacity deficits, 67
  conflict-and-peace divide, 66
  criminality in, 66
  economic undercurrents of
    criminality, 67
  effects of oil value chain, 144
  globalization of economic
    production, 84
  insecurity and violence, 84
  measure of capacity, 84

natural resource criminality, 86–7
oil sector criminality, 75
poverty and material
    deprivation, 84
resource curse, 210–11
resource management, 8, 211
timber, 69–71
weak capacity for official, 84
Voluntary Partnership Agreement
    (VPA), 72
vulnerabilities, 23–5

Washington Consensus, 19, 229
water-wars thesis, 121
West African Commission on
    Drugs, 231
World Economic Forums, 228
world systems theory, 105–6, 108
World Trade Organization (WTO), 97

Zambia
  dependency on one
    commodity, 100
  oil or mining legislation, 4
  partnership with China, 52
  socialist path to development, 17